Innovative Approaches to Global Sustainability

Edited by Charles Wankel and James A. F. Stoner

in collaboration with Shaun K. Malleck, Matt Marovich,
Susan Nave, and Jurate Stanaityte

INNOVATIVE APPROACHES TO GLOBAL SUSTAINABILITY
Copyright © Charles Wankel and James A.F. Stoner, 2008.

First published in hardcover in 2008 by PALGRAVE MACMILLAN® in the
United States – a division of St. Martin's Press LLC, 175 Fifth Avenue,
New York, NY 10010.

Where this book is distributed in the UK, Europe and the rest of the world,
this is by Palgrave Macmillan, a division of Macmillan Publishers Limited,
registered in England, company number 785998, of Houndmills, Basingstoke,
Hampshire RG21 6XS.

Palgrave Macmillan is the global academic imprint of the above companies
and has companies and representatives throughout the world.

Palgrave® and Macmillan® are registered trademarks in the United States,
the United Kingdom, Europe and other countries.

ISBN: 978-0-230-10405-1

Library of Congress Cataloging-in-Publication Data

Innovative approaches to global sustainability / edited by Charles Wankel
and James A.F. Stoner.
     p. cm.
ISBN 978-0-230-60804-7
1. Sustainable development. I. Wankel, Charles. II. Stoner,
James Arthur Finch, 1935–

HC79.E5I51494 2008
658.4'08—dc22   2008011329

A catalogue record of the book is available from the British Library.

Design by MPS Limited, A Macmillan Company

First PALGRAVE MACMILLAN paperback edition: April 2010

10 9 8 7 6 5 4 3 2 1

Printed in the United States of America.

Transferred to Digital Printing 2010

Innovative Approaches to Global Sustainability

Contents

List of Figures

List of Tables

PART I

Exploring Possibilities for Change

CHAPTER 1

Introduction: Exploring New Frameworks, Practices, and Initiatives for a Sustainable World

James A. F. Stoner and Charles Wankel

In *Experiments Never Fail* (2008), his update of *The Max Strategy* (1996), Dale Dauten reminds us how important it is to keep trying new things—day after day and time after time. Jim Collins and Jerry Porras learned from 3M and reported in *Built to Last* (1997, p. 6), "What looks *in retrospect* like brilliant foresight and preplanning was often the result of 'Let's try a lot of stuff and keep what works.'" Dauten (2008) would have us go beyond problems to the antiproblem and beyond failures to the antifailure.

We are now, all of us, all six-and-a-half billion of us, at a time in personkind's history where trying new things is a prerequisite for the survival of our species and for the survival of as many as possible of those species mentioned in *Fighting for Love in the Century of Extinction* (Goodstein 2007) that have not yet been lost. We are at a time where we are challenged to try new things and to learn from these experiments at an unprecedented, but hopefully not at an impossible, rate. The contributors to this volume offer a variety of innovative approaches to global sustainability—some may turn out to be useful and others may not. But all the approaches have one thing in common: a recognition that creating, trying out, rejecting, or modifying and improving, and trying again to find new ways of moving this species toward a socially just and ecologically sustainable way to occupy this planet is the only game big enough for any of us to play.

This volume will be only one of a great many recent-past and near-future volumes seeking innovative ways to transform the ways we work, live, consume, and experience our lives. It offers the chance to contribute in some way to assisting the generation of students now emerging from college who must stand up and be, as Eben Goodstein called upon them to be at the national Focus the Nation (http://www.focusthenation.org/) event on January 31, 2008, the "Greatest Generation"—the generation that will lead the world to end its dependence on carbon and move to a carbon-free and sustainable economy and society.

With the chapters in this book, and the innovative approaches contained therein, let the games continue. And let the playing fields be businesses, zoos, and universities; let the actors be companies in the Canadian oil and gas industry, college professors, and sustainability coordinators; let the ideas be green chemistry, EVA (economic value added), SOAR (Strengths, Opportunities, Aspirations, and Results), and triple bottom lines; and let the results be our learning to "meet this generation's needs in ways that *enhance* the capacity of future generations to meet theirs" in "a world that works for everyone with no one left out" (Stoner, 2006).

In Chapter 2, "'SOAR' from the Mediocrity of Status Quo to the Heights of Global Sustainability," Jacqueline M. Stavros and Joseph R. Sprangel Jr. start the search for innovative approaches to global sustainability by describing the SOAR model of organizational change and transformation and by reporting on how it is being used to speed up the process of aligning organizations with the need for global sustainability. In doing so, they take another step in the ongoing development of "Appreciative Inquiry" and add to the set of effective approaches that managers and administrators can use on this journey toward a sustainable world. Building on more than two decades of research, teaching, transformation initiatives, and writing provided by David Cooperrider and his colleagues (Srivastva and Cooperrider 1990; Cooperrider, Whitney, and Stavros 2003; Cameron, Dutton, and Quinn 2003), Stavros and Sprangel bring the Appreciative Inquiry approach to the most pressing issue of the world today—the search for ways to preserve the future of our species and all other species. Their chapter presents a business case for efforts to become a sustainable organization and gives examples of the financial successes of companies that have made this commitment. They also prove the consistency of SOAR with the literature of strategic management and with the current thinking of some leading management scholars.

To achieve these ends, the chapter first lays out a business case for sustainable development and profiles the strides E. I. du Pont de Nemours and Company has made through its three-pronged strategy, led by chairman and chief executive officer (CEO), Chad Holliday Jr. Next, the authors lay out the possible issues—stakeholder pressures, the impact of managerial conflict, the CEO and top management's impact on change, and the role of decision making—that organizations face when attempting to shift from status quo to a sustainability approach. Reviews of recent work in the field of strategic management, such as Griffiths and Petrick's (2001) review of corporate architecture alternatives for organizations interested in sustainable development, and Hart and Milstein's (2003) key dimensions of a stakeholder-value framework are examined in the third section of this chapter. They describe SOAR as a strategic-management framework that is beginning to "lead organizations to a path of sustainable development and transformational change by engaging the whole system into the approach." They then conclude the chapter by profiling some organizations that have used SOAR to achieve their sustainability-related goals. The clearly stated steps of the SOAR model and the substantial literature now available on Appreciative Inquiry that supports the model should be especially valuable to managers and others committed to moving their organizations off the "stuck status" of being unable to make a commitment to and to take action for the future.

In Chapter 3, "Sustainable Wealth Creation beyond Shareholder Value," Mats A. Lundqvist and Karen Williams Middleton outline what Sweden's Chalmers School of Entrepreneurship and the Göteborg International Bioscience School have done to encourage and prepare young entrepreneurs to follow a path toward building companies that contribute to global sustainability. And they report that the efforts have met with a remarkable level of success. The initial activities began in 1997 and have developed into a well-thought-out program consisting of an interwoven network involving students, educators, advisors, alumni, accountants, and lawyers—all collaborating on entrepreneurial ideas that are developed and nurtured by the students as they strive to establish financially successful companies that contribute to a sustainable world. An array of case studies are presented in this chapter, including the role played by a student, Anna Weiner, in ELLEN AB, a biotech company focused on women's health. While not all companies originating in this program succeed, this method of teaching, which combines education and venture creation, is a key takeaway for students, and the spirit of an "entrepreneurial ecosystem" remains with them in future pursuits.

Chapters 4 and 5 both address the very broad, multilayered, complex, controversial, and important issue-domain of measurement in the *problematique* of global sustainability. The two chapters bite off and chew on the closely related issues of measuring the impact of companies on the world and measuring the impact of green initiatives, commitments, and strategies within companies on the economic health of those organizations.

In Chapter 4, "Limits of Shareholder Value to Achieving Global Sustainability," Frank Figge and Tobias Hahn explore the question of whether or not the goal of sustainable development is in line with the goal of maximizing shareholder value—do companies "do well by doing good" or is there a limit to "how much a company can afford to 'spend' on sustainability." In doing so, they take a measured approach to analyzing the link between environmental, social, and economic performance in corporations. To explore this broad question, they present a method of measuring corporate-sustainability performance in monetary terms and apply it in detail to a specific company's operations and more broadly to a number of companies in the automobile industry (BMW, Daihatsu, GM, and Renault), calculating where each falls on a Sustainable Value matrix. The analyses provide a clear argument for the conclusion that sustainability "and shareholder value can be in line, but that there is no unambiguous link between the two." The analyses provide examples of "companies that create shareholder value and contribute positively to sustainable development" and of "companies that create shareholder value at the expense of sustainable development." Figge and Hahn indicate that the approach they present offers a new way of perceiving ultimate shareholder value and that the assessment of corporate sustainability performance will start playing a dominant role in this equation. As they round out their chapter, they conclude that it is "clear that companies need not only economic capital but also environmental and social resources to create a return" and they discuss these implications and challenges.

In Chapter 5, "Green Chemistry and EVA: A Framework for Incorporating Environmental Action into Financial Analysis," Geoff Archer, Andrea Larson, Mark White, and Jeffrey G. York introduce "green chemistry" as an approach that involves, among other things, reducing "the amount of materials employed, and particularly the levels of hazardous substances used or emitted in the production of manufactured goods." They argue that managers may not need to adopt new financial tools such as the triple bottom line, which we will see supported elsewhere in this book, but might instead adopt a new tool that combines the twelve principles of green chemistry. They explain each of the principles, provide examples of each, and assess the relationship between each principle and the popular financial metric known as EVA—a tool long used to evaluate investment proposals. These twelve principles, once integrated, will create what the authors believe is an effective method for companies to evaluate the costs and benefits of sustainable business decisions in an environment that is challenging them to find success in both financial and sustainable strategies.

In both the title of Chapter 6, "Monkey See, Monkey Do?" and in the content, Nicole A. M. Horstman, Frank G. A. de Bakker, Enno Masurel, and

Patricia P. van Hemert bring a fresh and unusual theme onto the playing field of global sustainability—a place too often mired in the state of inaction created by the "Twin D's of Global Doom: Denial and Despair." They focus on innovations for sustainability in Dutch zoos, institutions that generally have been overlooked when seeking solutions to problems in industry or other sectors of the economy. Zoos, with their multifaceted missions related to biodiversity, environmental education, preservation, research, and preserving endangered species, can serve as models of knowledge exchange and resource dependence in industry and elsewhere. The authors stress that innovation is a long-term activity that needs research and finance to support it and that it is usually the larger and more financially booming companies and industries that succeed in innovation. Although each is of a small scale, individual zoos share knowledge to create a network of zoos (among countries and even worldwide) that yields a large pool of resources for each to contribute to and take from, thus helping each zoo develop more sustainable innovations and increasing the total pool of innovations for all. Zoos have come to adopt a process where other zoos are their biggest allies, and by sharing knowledge, they further enhance the success of all zoos and ultimately help accomplish what they are all there for: the well-being of the animals and of society.

The authors emphasize the zoos' recognition of their shared fate, even as they compete with one another for audiences and resources and simultaneously collaborate in innovation and learning. As stated by one zoo manager who took part in the research, "If things go wrong in Dutch zoos, the audience will remember; this will affect our zoo as well, our image." Those words, reflecting the wisdom of the zoo managers and their institutions, recall the similar wisdom of industry leaders who have begun to observe that "there are no profits to be made on a dead planet." The authors conclude by building on the research interviews. They report to offer a five-step process on how organizations in any industry can embark on one of the key skills required for them to contribute to a sustainable world—learning to work with their competitors to develop and share sustainable innovations more quickly and successfully.

In Chapter 7, titled "Toward Environmental Sustainability: Developing Thinking and Acting Capacity within the Oil and Gas Industry," Laurie P. Milton, and James A. F. Stoner suggest a bold new role for the Canadian oil and gas industry: recreate itself to become a sustainable-development model for all industries. They state that the process of providing such leadership can be anchored in what we are learning about the best practices in achieving transformational change at the individual, organizational, and societal levels. In calling upon Canadian companies—individually and collectively—to accept this challenge, they identify six foundations of such change that the companies and industry can build upon—creating an inspiring goal, engaging in heedful collective thinking and action, engaging in resilient thinking and action, building intercompany collaboration, developing inspired and inspiring leadership, and taking actions directed at getting society on board.

In their discussion of this possible global leadership role for a Canadian industry, they address the question of why the Canadian oil and gas industry might be an exceptionally promising candidate to provide such leadership and provide examples of instances in which members of the industry are already taking such actions. Because of the key importance of cooperative and collaborative thinking with other firms in the same industry, they pay particular attention to the research-based evidence and theory underlying our knowledge of the conditions that make such thinking and actions possible and successful. And, of course, they stress the role inspiring leadership can play in such a bold undertaking. In line with the growing importance and value of studies in the realm of positive psychology and positive organizational scholarship, they stress the importance of such companies remaining not only positive but also realistic at the same time, in their pursuit of leadership roles in creating a sustainable world.

Paul Shrivastava, Douglas E. Allen, and Tammy Bunn Hiller's chapter on "Designing Undergraduate Education on Managing for Sustainability" is based on Bucknell University's experience in building a sustainability-management program from the ground up: identifying sustainability-management concepts, skills, and tools and then developing an educational program to deliver them. They describe developing undergraduate education in managing for sustainability in the context of a liberal arts university. The program was developed with a clear grounding in globalization, transcending national boundaries and national regulatory bodies. Here globalization is characterized by free and rapid investment-capital flowing from one region of the world to another, operations shifting somewhat quickly from region to region based on the cheapest labor rates, and global supply chains transporting food from distant locales with consumers having little information about its origin or the conditions in which it was grown; in short, it is postmodern capitalism where corporations have tremendous powers. Their program presents a situation where corporations have assimilated environmental and related social concerns. Interactions between corporations and environmentalists are presented as no longer being mediated by governmental agencies. Organizations' system-wide impacts and roles are seen as requiring a systemic view of organizational sustainability, including not only economic and ecological performance but also social and ethical performance. They address the concept of the "triple bottom line" (Savitz and Weber 2006) as one way of understanding corporate sustainability. At Bucknell, the faculty members acknowledge that corporations must earn a reasonable return for shareholders and they are committed to helping students learn to balance these goals with concerns for social and environmental justice. Knowledge relevant to sustainability is spread across the university. Bucknell shows how faculty, resources, and administrative attention can come together to create a sustainability-management program.

In Chapter 9, "The Sustainability Coordinator: A Structural Innovation for Managing Sustainability," Gordon Rands, Barbara Ribbens, and David Connelly report on the emergence of a new organizational role that shows promise in speeding up the movement of organizations toward a more sustainable way of being. They add their voices to the other authors in this volume who stress how important it is to act now to create ecologically, socially, culturally, and economically sustainable societies, and they report on the promise that the position of "sustainability coordinator" holds for building the necessary commitments to global sustainability and speeding up the processes of taking action to make it happen. They report successes at the academic level with key collaborations among students and academics at a number of universities such as UCLA, Yale, and Harvard. Perhaps influenced in part by successes at the university level, interest in this role has been growing in the public sector, with programs being conducted across the country from California to Colorado to Ohio. The authors profile what has been going on in the private sector, with companies such as Ben and Jerry's and Office Depot, although these authors feel such programs are more the exception than the rule in the private sector. They attribute the slowness of the private sector in adopting this type of innovation to a number of factors—especially to stakeholder and economic pressures. However, they and, we believe, all of the authors in this volume recognize and are encouraged by the fact that "The Times, They Are A-Changin."

We wish you the best of luck and much courage as you try new ways of creating "a world that works for everyone with no one left out."

Acknowledgments

We are more than a little grateful to Mathew Marovich, Susan Nave, and Jurate Stanaityte who played major roles in making this chapter and all of this book happen. Thank you.

References

Cameron, K. S., Dutton, J. E., and Quinn, R. E. 2003. *Positive Organizational Scholarship: Foundations of a New Discipline.* San Francisco, CA: Berrett-Koehler.

Collins, J., and Porras, J. 1997. *Built to Last: Successful Habits of Visionary Companies.* New York: HarperBusiness.

Cooperrider, D. L., Whitney, D., and Stavros, J. M. 2003. *Appreciative Inquiry Handbook.* Bedford Heights, OH: Lakeshore.

Dauten, D. 1996. *The Max Strategy.* New York: Morrow.

Dauten, D. 2008. *Experiments Never Fail: A Guide for the Bored, Unappreciated and Underpaid.* Richmond, CA: Maurice Bassett.

Goodstein, E. 2007. *Fighting for Love in the Century of Extinction: How Passion and Politics Can Stop Global Warming.* Lebanon, NH: University Press of New England.

Griffiths, A., and Petrick, J. A. 2001. Corporate Architecture for Sustainability. *International Journal of Operations and Production Management* 21 (12): 1573–1586.

Hart, S. L., and Milstein, M. B. 2003. Creating Sustainable Value. *Academy of Management Executive* 17 (2): 56–69.

Savitz, A., and Weber, K. n2006. *The Triple Bottom Line.* New York: John Wiley.

Srivastva, S., and Cooperrider, D. L. 1990. *Appreciative Management and Leadership: The Power of Positive Thought and Action in Organizations.* San Francisco: Jossey-Bass.

Stoner, J. A. F. 2006. It's Not About the Profits. Keynote address, Third International Research Conference on Business Management. *Corporate Social Responsibility: Profits and Beyond.* University of Sri Jayewardenepura, Gangodawila, Nugegoda, Sri Lanka.

CHAPTER 2

"SOAR" from the Mediocrity of Status Quo to the Heights of Global Sustainability

Jacqueline M. Stavros and Joseph R. Sprangel Jr.

Introduction

A letter written by Walter Isaacson, the president of the Aspen Institute, a nongovernmental organization (NGO), to Aspen friends and partners reads

> At certain points in our lives, many of us feel the need to reflect on what it takes to lead a life that is good, useful, worthy, and meaningful. Perhaps we have noticed ourselves trimming our principles and making too many compromises in our careers, and we want to reconnect with our values. Or perhaps we yearn, in a world filled with clashing opinions, to understand the great ideas and ideals that have competed throughout the progress of civilization.
>
> (Isaacson 2006)

Some people, as they reach a certain stage in their life, look at themselves in the mirror and ask, "How can I make a meaningful and sustainable contribution?" This question could be thrown up by some event in their life, such as a midlife crisis, frustration with a current job, or introduction to a new concept. Ray Anderson, the former president of Interface Inc., had a similar moment. He knew there was a greater sense of purpose a leader could pursue beyond merely seeking to maximize shareholder wealth.

In 1994, Interface, an international carpet manufacturer, embarked on a journey to become a company whose operations were fully aligned with the needs of a sustainable world and to demonstrate how other companies could succeed in the same way. Today Interface is generally recognized as a leader in industrial sustainability. The company's Interface Sustainability website (Interface Sustainability 2006) lists the "Seven Faces of Mt. Sustainability." These include (1) waste elimination, (2) benign emissions, (3) renewable energy, (4) resource recycling, (5) resource-efficient transportation, (6) stakeholder sensitization, and (7) redesign of commerce. While waste elimination sounds like a great idea that is likely to increase company profits, one might see the rest of the initiatives as bearing significant costs that cannot immediately lead to an immediately improved shareholder return. However, some organizational leaders are beginning to see that sustainable-development pioneers such as Anderson have achieved financial as well as environmental and societal success with this strategic approach.

The Center for Business as an Agent of World Benefit (BAWB) website (BAWB 2005) provides several examples of companies whose sustainable-development initiatives have yielded financial successes. These companies include Green Mountain Coffee Roasters, the Body Shop, Tom's of Maine, Toyota Motor Corporation, Patagonia, Wal-Mart, and Unilever.

> With growing sensitivity toward social issues, companies are increasingly expected to take greater responsibility for making sustainable development a reality. However, defining this new role is a major challenge for companies as they search for ways to balance economic, environmental, and social performance. To integrate sustainability principles into their business strategies and to aid resource allocation decisions, managers must quantify the link between social and environmental actions and financial performance.
>
> (BAWB 2005)

This chapter discusses, in separate sections, four areas intended to help organizational leaders determine, in part, what it means to move from the current status quo of mediocrity to the heights of global sustainability. In the first section, a brief overview of sustainable development and a proposed business case for such development provide the reader with an understanding of this concept. That section concludes with a review of the work of an organization that has made the transition to become focused on sustainable development. In the second section, potential issues that organizations may face when trying to shift from status quo to a sustainable approach are shared. In the third section, a brief review of recent work in the area of strategic management is presented to suggest some of the factors that have contributed to the transformational change necessary to become a sustainable-development organization. In the last section, a strategic management framework titled

"SOAR" is introduced. The SOAR acronym refers to Strengths, Opportunities, Aspirations, and Results. SOAR is a flexible framework with a simple approach that invites the relevant *stakeholders* into the strategy creation or strategic planning process to create *sustainable value*. The design of this approach seeks to build upon the factors that enable strategic transformation to occur and is beginning to lead organizations to a path of sustainable-development and transformational change by engaging the whole system, or cross-sectional representatives of the system, in the process.

The Concept of Sustainable Development

The 1987 Brundtland Commission Report introduced the term "ecologically sustainable development," and this phrase was further cemented at the 1992 Earth Summit in Rio de Janeiro (Shrivastava 1995). The report presented a framework and a set of principles to address ways to protect our planet's resources while taking into consideration economic and social concerns. Over the last two decades, there has been a growing need to understand what *sustainable development* or *sustainability* means to an organization and to its business model to help solve the mystery of how to create a balance between profitability to shareholders and responsibility to stakeholders.

Building on the work of the Brundtland Commission, the World Business Council for Sustainable Development (WBCSD) defines sustainable development as "forms of progress that meet the needs of the present without compromising the ability of future generations to meet their needs" (WBCSD 2007). On both philosophical and practical perspectives, numerous authors concerned with issues of sustainability have challenged the traditional single-bottom-line profit measure of business success and the model of business that states that the sole purpose of the for-profit corporation is to maximize shareholder wealth (BAWB 2005; Sustainable Development Commission 2006; WBCSD 2007).

On many fronts, business and nonbusiness organizations are being called upon to take responsibility for a growing number of problems ranging from global warming and ozone depletion to issues of sustainable development. Referring to the future agenda of the leading international association of management scholars and teachers, Gladwin, Kennelly, and Krause state, "Transforming management theory and practice so they positively contribute to sustainable development is, in our view, the greatest challenge facing the Academy of Management" (Gladwin, Kennelly, and Krause 1995, p. 900). BAWB has been described as seeking to "[unite] the best in business with the call of our times . . . putting their people, imagination and assets to work to benefit the earth, from its ecosystem to the needs of its vast, diverse population" (BAWB 2005). The WBCSD website (WBCSD 2007) calls for a framework that supports a collaborative approach to individual and business-community action leading to sustainable development. The new premise is simple: "Corporations, because they are the dominant institution on the planet, must squarely address the social and environmental problems that afflict humankind" (Hawken 1993, p. xiii).

And there are warnings to businesses that do not heed these calls: "Companies that continue to approach environmental problems with band-aid solutions and quick fixes will ultimately find themselves at a competitive disadvantage" (Dechant and Altman 1994, p. 7).

The remainder of this section offers a proposed business case for the adoption of a sustainable-development approach to business and reports on the success of a sustainable-development proponent in efforts to adopt such an approach in an organization.

The Business Case for Sustainable Development

Organizations considering making the shift from status quo to sustainability need to understand whether there is a viable business case for doing so. A business case for sustainable development was described by WBCSD as follows:

> Pursuing a mission of sustainable development can make our firms more competitive, more resilient to shocks, nimbler in a fast-changing world and more likely to attract and hold customers and the best employees. It can also make them more at ease with regulators, banks, insurers and financial markets. Sustainable development policies will be profitable, but our rationale is not based solely on financial returns. Companies comprise, are led by, and serve people with vision and values. In the long-term, companies that do not reflect these people's best vision and values in their actions will wither in the marketplace.
>
> (WBCSD 2007)

The WBCSD is one of the organizations that can assist business leaders in developing their companies' own business cases for sustainable development and in moving forward to translate those resulting business cases into reality. It is made up of 180 international organizations committed to the following mission: "To provide organization leadership as a catalyst for change toward sustainable development, and to support the organization license to operate, innovate and grow in a world increasingly shaped by sustainable development issues" (WBCSD 2007). Their objectives include

- organization leadership—to be a leading organizational advocate on sustainable development;

- policy development—to help develop policies that create framework conditions for the organizational contribution to sustainable development;
- the organization case—to develop and promote the organization case for sustainable development;
- best practice—to demonstrate the organizational contribution to sustainable development and share best practices among members; and
- global outreach—to contribute to a sustainable future for developing nations and nations in transition.

Leaders might consider these elements when making a strategic transformational change toward bringing their organizations on the path of sustainable development. Because many executives may see merit in the objectives of sustainable development but wonder if it is possible to actually achieve this transformation in their organizations, reviewing the methods adopted by an organization that has embarked on this transition may be necessary.

A Sustainable Development Success Story

E. I. du Pont de Nemours and Company, led by Chad Holliday Jr., chairman and chief executive officer (CEO), is one example of an organization that has taken a proactive approach to sustainable development and has thoroughly embraced the concept (Holliday 2001). In addition to his work with DuPont, Holliday is the chairman of the WBCSD. As the leader of the DuPont organization, he realized that two fundamentals would drive the company's actions: (1) cheap supplies of hydrocarbons and other nonrenewable resources necessary for DuPont's products would not continue to be available and (2) the ecosystem has a limited capacity to absorb the emissions and wastes from the products and processes of DuPont and all other companies. Although he recognizes that these factors comprise significant dangers for a company such as DuPont and for the world, Holliday is also an optimist about the future: "Obviously we have only one Earth, and some argue that the standard of living in the developed world must fall for the developing countries to progress. I strongly disagree with this zero-sum mentality" (Holliday 2001).

Holliday has been pursuing a commitment to utilize creativity and scientific advances that achieve a strong return for shareholders while operating in a manner focused on sustainable development. The commitment to sustainable development does not exist only for altruistic reasons. It exists because the company is able to adhere to this commitment while meeting the needs of the organization and its shareholders. This philosophy goes against the beliefs and arguments of many organizations that they cannot afford to be sustainable. If DuPont is able to achieve this outcome, how does it do so?

The prerequisites for the success of DuPont's sustainable-development strategy are the alignment of that strategy with what the organization and its members care about and how the organization creates economic and societal value. The integration of scientific disciplines is strongly valued at DuPont, and the company's accomplishments in doing so have been key factors in its business success. DuPont achieves success in sustainable development through a three-pronged strategy that includes (1) integrated science, (2) knowledge intensity, and (3) productivity improvement. The company uses integrated science to develop products and processes that unite chemistry, biology, and other sciences. Knowledge intensity requires that knowledge content is added to products and processes in ways that create value. DuPont's work in the area of productivity improvement has elevated efficiency gains from an operational level to a strategic level. In pursing this three-pronged approach, sustainable growth is viewed by the organization as a comprehensive way of doing business and not as an environmental initiative. This approach has enabled DuPont to create shareholder value with processes that are increasingly consistent with the needs of a sustainable world and that yield products with the same characteristics. The approach has led to a vast array of new opportunities and increases in economic value for the organization. DuPont represents a success story of an organization focused on sustainable development and financial success that has been following its strategy with what Holliday describes as a "relentless determination and tenacity" (Holliday 2001).

The intents of this section have been to give brief overviews of the concept of sustainable development, the business case for pursuing it, and the success of one organization in doing so. We now provide a review of some of the issues an organization could potentially face in transitioning to a sustainable-development approach.

Issues that Support the Pull off Status Quo

Moving an organization off status quo requires an understanding of the factors that frequently keep desired changes from occurring. Four common factors are stakeholder pressures, the impact of managerial conflict, CEO and top-management impacts on change initiatives, and the role of decision making.

Stakeholder Pressures

Waddock, Bodwell, and Graves (2002) have identified three sets of pressures that organizations currently face. The first set of pressures comes from the primary stakeholders that include owners, suppliers, employees, and customers.

The second set of pressures is felt from secondary stakeholders that include NGOs, governmental regulatory agencies, community groups, and special interest groups. Recently, a third set of pressures has been brought on by a change in social and institutional norms that, in turn, have yielded an influx of "best of" rankings and global reporting standards that assess the economic, social, and environmental performance of organizations. To deal with these pressures, organizations must use a strategic approach that includes inquiring into and discovering the diverse requirements of these various stakeholders. The approach must also include a plan and a set of actions that leave all parties feeling that they have been genuinely heard, even if they have not been successful in getting all that they would have liked.

Waddock, Bodwell and Graves recommend four actions for organizations interested in dealing effectively with these pressure groups. First, create a vision statement and values set aligned with baseline global standards. Second, through an ongoing dialogue, communicate to these pressure groups what expectations they can hold for the organization's behavior based on those values and organizational vision. Third, integrate the strategies and management objectives of the organization into the actions of its employees. Fourth, develop and use key performance indicators to communicate organization performance against the defined global standards. These actions align with those that can be found in successful organizations and in the field of strategic management. The SOAR framework, introduced in this chapter, brings these four key areas into action.

In reporting his research, Frooman (1999) also argued for the importance of managers to understand the impact of stakeholders and the influence strategies various stakeholders were likely to use. It is important to include stakeholders' desires when determining the strategic plan of the organization. In this process, it is most important to develop environmental-improvement initiatives that are consistent with the pressures likely to be generated by various influential stakeholders and to balance the diverse expectations of those stakeholders with these initiatives.

In the context of stakeholder pressure and environmental performance, Kassinis and Vafeas (2006) report that the reason members of wealthy communities live in a cleaner environment than do members of poorer communities is that the wealthy have greater resources and power. Organizations are more dependent on and responsive to the more influential groups and are more likely to meet their demands. They also observed how proenvironmental pressures from organized groups at the state level can be successful in putting pressure on organizations to improve their environmental record. There was a finding that even though only 16 people out of 1,000 residents of one state were involved in environmental initiatives, the group improved things for all residents of the state. Their results also showed that the denser the population the more likely it was that pressure would be placed upon an organization to reduce emissions. An act as simple as writing a check to an environmental membership group helped to put pressure on the management team of an offending organization. As leaders of organizations develop sustainable-development actions, they will need to understand both the various stakeholder groups and the level of economic influence within each of these groups. Actions counter to the expectations of the wealthier members of a defined stakeholder group or of some emerging nonwealthy but very publicly visible community groups such as Sustainable South Bronx (Sustainable South Bronx 2008) may cause issues for organizations that do not understand the economic and political impact of these actions.

The Impact of Managerial Conflict

Changing from status quo to a focus on sustainable development will likely create issues of managerial conflict in an organization. Dealing with this conflict in a positive manner is another key element in the transition to sustainability.

Amason (1996) found that conflict can help managers make effective decisions, but the nature of the conflict is very important to the results it yields. He found a distinction between the positive impacts of cognitive conflict as opposed to the injurious impact of affective conflict. Earlier research had called attention to a similar possibility. Schweiger, Sandberg, and Ragan (1986) noted that "on the one hand, conflict improves decision quality; on the other, it may weaken the ability of the group to work together." Amason did not find that using conflict as the driving force yielded higher-quality decisions. But he did find that the quality of decisions was improved with the input of multiple perspectives from stakeholders. Additionally, other research he reviewed suggested that team members are more likely to accept and support decisions that are open and considerate of their input. He concluded that two types of conflict exist and must be addressed. The use of cognitive conflict should be encouraged and that of affective conflict discouraged. When conflict is addressed in this manner, the end result should be greater acceptance of high-quality decisions. The SOAR approach recognizes that conflict may have some benefit in decision making but prefers an avenue of dialogue as the first course of action.

CEO and Top-Management Impact on Change

Organizations that maintain the status quo will tend to be entrenched in their ways of being and behaving for various reasons. One cause is the impact of the makeup and length of service of the top-management team. Organizational leaders desiring to transition to a sustainable-development focus would do well to understand the current executive groups in their organizations. A study by Boeker looked at the strategic-change impact of the CEO and other top managers (Boeker 1997). The study emphasized top management's direct and interactive initiatives to influence performance and the resulting performance of the organization. Boeker looked at (1) chief executive succession, (2) chief executive tenure, (3) top management tenure, (4) top management diversity, and (5) the interactive effect of these group characteristics on strategic change. According to Boeker:

- poorly performing organizations had a misalignment between their strategy and business environment, leading to a need for strategic change.
- there was an independent influence by top management on organizational decisions pertaining to strategy direction.
- the tenure of the CEO had a significant impact on the strategic change of the organization, but such a relationship was not found for CEO succession.
- the tenure, tenure diversity, and homogeneity of the top management team were also found to have an impact on the extent of strategic change.
- organizations with long-tenure CEOs and top-management groups were less likely to adopt strategic-change initiatives.
- large public organizations were more likely to initiate strategic change than smaller and privately held organizations.

As organizational members prepare to develop the strategic plan for their organizations, it would be wise to remember the makeup of the top-management team. If the team has been in place a long time, and the tenure of the group is similar, they should reflect on this fact to see if it has caused them to be overly conservative or complacent in their approach to planning.

The Role of Decision Making

Twenty years of research by Paul C. Nutt shows that organizations fail with half of their decisions (Nutt 1999). He found three common mistakes made in the decision process used by organizations. These include (1) effective practices are commonly understood yet not typically used; (2) decision makers take shortcuts because of perceived pressure; and (3) effective problem-solving environments are not created. The research showed a common understanding of the need for employee participation, yet this method was used in only one out of five decisions. The rationale for taking shortcuts was the desire to copy the practices of successful organizations, but the attempted shortcuts involved a lack of understanding of the difficulties that would be faced in implementing the change. A preferred approach to taking shortcuts is to ask people to find the root cause of an issue, which leads to a focus on placing blame when asking them to find opportunities or solutions to a problem and which also places an emphasis on results. SOAR focuses on identifying the strengths, opportunities, and aspirations that lead to measurable results involving employees and stakeholders.

Nutt's research led him to define six elements managers and others can use in making effective decisions and then in implementing them successfully. These elements are (1) to participate actively in decision making and to act as champions for the resulting decisions; (2) to set an environment of understanding the real issues that must be faced in making the planned implementation a success; (3) to follow up on the need for interventions by setting clear directions and objectives for the interventions; (4) to focus on developing the creative ideas that are needed for effective implementation; (5) to understand the importance of developing alternative options that allow for dialogue on the merits of the various options, which should result in a consensus selection of the best implementation options or a hybrid of several alternatives; and (6) to realize that in the end none of these elements can succeed if the manager does not remove barriers faced by the team involved in the process of developing and implementing the defined solution. We see these recommendations as key factors to utilize in the implementation of sustainable-development initiatives.

Further research has emphasized that the effectiveness of the processes used to make key decisions for an organization are in the hands of its managers (Astley and Van de Ven 1983; Dean and Sharfman 1996; Hitt and Tyler 1991). Managers need to understand the importance of their decision-making processes when determining initiatives to improve competitiveness that impact the survival of the organization. Dean and Sharfman found that there was a favorable influence on the effectiveness of key decisions in stable environments (Dean and Sharfman 1996, p. 389). This is not apparent on those made in unstable environments. From our perspective, this supports that it would be rational to assume that environmental instability would render the likelihood of any decision having the chance to be successful.

As presented earlier in this chapter, the decision to be proactive in creating environmental improvement initiatives can lead to effective integration of these initiatives into the overall strategic plans for the organization. DuPont and Interface have been effective in doing so. These organizations have found that integrating sustainable development into the strategic-planning actions of the organization can lead to an improved organization climate and improved business results.

Creation and Development of Strategy

Now that the concepts of sustainable development and the issues an organization may face in developing a new strategic focus have been introduced, the issue of creating and developing a strategy focused on sustainable development will be addressed. Wheelen and Hunger define *strategic management* as "that set of managerial decisions and actions that determines the long-run performance of a corporation" (Wheelen and Hunger 2006, p. 3). This section addresses four aspects of the task of creating and implementing a strategy focused on sustainable development by (1) presenting a strategic approach to ecological improvement; (2) introducing a sustainable enterprise framework; (3) reviewing corporate architecture alternatives that can support a proactive strategy like managing for sustainable development; and (4) reporting some of the perspectives of leading strategic management scholars on the likely future of the process of strategy creation and development.

Strategic Approaches to Ecological Improvement

Shrivastava states that "ecological problems rooted in organizational activities have increased significantly, yet the role organizations play in achieving environmental sustainability is poorly understood" (1995, p. 936). He observes that organizations have tremendous power to initiate significant ecological improvement with the resources, knowledge, and power they possess. The incentives for organizations to support ecologically sustainable development activities include (1) driving down operating costs; (2) creating a competitive advantage with green consumers; (3) ecologically establishing leadership in their industry; (4) establishing a legitimate sustainability presence with the public and stock markets; (5) reducing long-term risks associated with resource depletion; (6) improving ecosystem and community environment; and (7) positioning their organizations ahead of the regulatory curve. This list of possible benefits will not happen with a six-to-twelve-month payback. This approach requires up-front costs to be incurred and a long-term commitment to sustaining an organization.

Shrivastava's recommended strategy for initiating ecological improvement includes: (1) creation of green production systems; (2) encouraging the emergence of green markets through first-mover strategies; (3) efficient cost strategies that result in long-term ecological efficiencies; (4) improved legal systems that promote ecological product liability reductions; and (5) the development of true green programs leading to improved public and community relationships. He recommends that these improvements begin with simpler actions and build upon early successes to advance to more significant initiatives at a later date. This progression appears to be a logical framework to follow when organizations are interested in becoming more sustainable.

A Sustainable Enterprise Framework

A framework an organization can utilize to become a sustainable enterprise is presented by Hart and Milstein (2003). The framework is a multidimensional construct used to create sustainable value for the organization. The vertical axis shows the needs of today at the bottom end and the needs of tomorrow at the top end. The horizontal axis has the needs of the internal stakeholder at the left end and the external stakeholders at the right end. The intersection of the X and Y axes is where the extremes denoted by the end of each axis are in balance. The upper-left quadrant is the strategy of clean technology where the organization develops the sustainable competencies of the future, leading to a corporate payoff of innovation and repositioning. The upper-right quadrant is the strategy of a sustainable vision that leads to a roadmap for meeting unmet needs with a corporate payoff of growth trajectory. The lower-left quadrant is the strategy of pollution prevention to minimize waste and emissions from operations leading to a corporate payoff of cost and risk reduction. The lower-right quadrant is the strategy of product stewardship to integrate stakeholder views into organization processes, leading to a corporate payoff of reputation and legitimacy.

Hart and Milstein feel that sustainable value creation is a huge and yet largely untapped opportunity for organizations. They see the framework as a means of establishing an understanding of the nature and magnitude of the introduction of sustainable-development initiatives. While the framework is simple, the work to achieve the stakeholder balance is complex and difficult to accomplish. The SOAR framework provides a process for bringing stakeholders into the strategy-making and implementation process to achieve this desired balance of sustainable-development value.

Corporate Architecture Alternatives and Proactive Strategy

Griffiths and Petrick (2001) addressed the subject of corporate-architecture alternatives for organizations interested in sustainable development. They first reviewed prior research that investigated three types of resistance facing organizations seeking to become sustainable enterprises. First, organizations tend to insulate themselves from available environmental information. Second, organizations establish routines that support the maintenance of status quo. Third, organizational architectures often limit the range of stakeholders who participate in the planning process. These three factors tend to lead to reactive rather than proactive sustainable-development actions.

Griffiths and Petrick's review of corporate architecture included alternatives to standard approaches. These were (1) network organizations, (2) virtual organizations, and (3) communities of practice. Network organizations allow small constituent groups to be part of a larger organization. The small groups promote sustainability because they are flexible, are able to respond quickly, and can experiment with new innovations on a small scale. As markets grow, network organizations can duplicate pieces of themselves to take on increased business. Virtual organizations can have a limited life that is defined to meet a specific need for a set period of time. They can promote sustainable development because they exist only while necessary and can be easily disbanded when they are no longer needed. Communities of practice create an environment of innovation where tacit and explicit knowledge can be shared and developed into sustainable-development practices to improve formal organizations.

The three alternative architectures described by Griffiths and Petrick can generate practices that help organizations break away from past architectures that have impeded improvement. Use of these alternative forms can also tap into the capabilities of the entire knowledge base of the organization, resulting in the opportunity for enhanced adaptation and rapid response to an organization's opportunities or threats. Griffiths and Petrick make clear that current architectures can inhibit an organization's chance of becoming capable of sustainable development. Being aware of this danger can encourage organizations to develop an architecture that is right for its needs. We agree with the need to develop alternative architectures to improve the capability of organizations to transform themselves into ones built on sustainable development. The involvement of key stakeholders to identify opportunities and collectively move the organization forward is a premise we strongly support.

Next we look at what some leading management scholars have said recently and in the past about the role of strategic management in helping organizations develop strategic-change actions. The expectation is that these actions will lead to transformational change. Transformational change takes into consideration the external environment, leadership, mission, vision, strategy, organizational culture, and overall organizational performance (Burke and Trhant 2000). It is our position that most organizations will not succeed in their attempts to embrace the concept of sustainable development unless they experience transformational change consistent with the insights presented next.

Influential Management Thinkers' Insights on Strategy

The Suntop Media Thinkers 2005 is a listing of the fifty most influential living management thinkers (Suntop Media Thinkers 50 2007). Those on the list include Gary Hamel, Constantinos Markides, Henry Mintzberg, and C. K. Prahalad. Their viewpoints validate a need for an alternative approach to strategy that accelerates strategic planning and brings in the viewpoints of an organization's stakeholders.

Hamel and Prahalad posed the following questions prior to the 15th Annual Meeting of the Strategic Management Society in Mexico City in 1995:

> How to create an organization that really, truly lives in the future, and then interprets today's decision in that context? How does one unleash corporate imagination? How does one turn technicians into dreamers? How does one turn planners into strategizers? Is there no recourse except to sit back and wait for the visionary to emerge? Planning may be discredited, and strategists on the run, but managers must not shirk from the responsibility of leading their organizations to the future.
>
> (Hamel and Prahalad 1996)

A recent strategy perspective was presented by Daniel McCarthy as an outcome of a crosstalk interview with Henry Mintzberg, who states:

> Certainly leaders make a difference. There is no question about it. But leaders often make a difference because they stimulate others, not because they come in with grand strategy. But what we're getting now, very dangerously, is what I call a dramatic style of managing: the great merger; the great downsizing; the massive, brilliant new strategy. Most of this is junk and fails utterly, but not until it fools the stock analysts for a few years.
>
> (McCarthy 2000)

In commenting on the McCarthy interview with Mintzberg, Markides noted that anyone in the organization can conceive new strategic ideas. The use of a democratic approach in which thousands of employees are asked to provide input to the strategic plan allows greater potential for innovative suggestions than does limiting the ideas to just those brought forward by a handful of senior executives. The role of determining which of these ideas are included in the strategic plan is the responsibility of top management.

According to de Kluyver and Pearce, "Only a CEO or senior management can drive the strategic thinking process . . . (which) through a series of exchange works its way down to involve every level of the organization" (de Kluyver and Pearce 2006). This task of bringing the strategic thinking process to all levels of the organization needs to be done effectively to keep chaos and confusion from occurring that ultimately could result in a disgruntled or demotivated employee base.

Later in the McCarthy interview Mintzberg describes the pendulum swing from the focus on competitive analysis and industry analysis back to dynamic capabilities. He speculates that it may be time to go back to the use of the SWOT (Strengths, Weaknesses, Opportunities, and Threats) approach (McCarthy 2000; Thompson and Strickland 2005). A SWOT analysis is a first step in the traditional approach to strategic management. Strategic planning is typically a top-down centralized process with a team assigned to determine the future of the corporation by assessing internal and external variables using a SWOT format. SWOT is a common strategic-planning tool that has served to maintain the status quo of business (de Kluyver and Pearce 2006). Might there be something that allows an organization to soar to new heights? We propose the use of an emerging framework that leverages, yet evolves beyond, SWOT.

The SOAR framework attempts to address the concerns raised above, in part, by building upon the possibilities offered by the strategy frameworks discussed briefly in the next section.

Strategic Frameworks

This review provides alternatives for leaders that address the questions and concerns presented earlier. First, a variety of strategic frameworks will be reviewed focusing on the strengths and weaknesses of the presented frameworks. Second, reviewing the practices of leaders, we will discuss the steps taken to move from ecological compliance to sustainability. Third, the impact of stakeholder theory on strategic management will be considered. Finally, concerns will be raised on the desire of stakeholders for fairness and participation in the development of strategic plans.

After reviewing the strengths and weaknesses of a variety of strategic frameworks and approaches used for competing in a globally competitive environment (which included the resource-based view, core competence, competitive advantage of nations, strategic groups, cognitive communities, network approaches, and competing for the future), Thomas, Pollack, and Gorman (1999) concluded that the frameworks are useful for creating multiple lenses with which to view the competitive environment.

Environmental leadership practices that move an organization from compliance to sustainable competitive advantage were reviewed by Dechant and Altman (1994). In looking at the increased pressures to become "green" that faced organizations, they listed environmental forces similar to the ones discussed earlier in this chapter. The forces they identified came from (1) an attempt to stay ahead of regulation, (2) stakeholder activism, (3) changing employee expectations of organization environmental performance, and (4) competitive pressures. In the article, they recommend "best green practices" for leadership to follow when interested in becoming a sustainable-development organization. The best-practice process starts with a mission statement and with corporate values aligned with environmental initiatives. These practices recommend that a shared vision be created as a subset of the values that support an environmental approach to conducting organization actions. An effective framework for managing this revised approach must be designed. The organization will also need to strengthen and revise product and process designs to align manufacturing operations with the new values, vision, mission, and strategic initiatives that have been put in place. Stakeholder partnerships should be developed to expedite transition to green practices. The final best practice involves the education of internal and external stakeholders on the need and actions that will lead to an organization based on sustainable development. And Dechant and Altman added their voice to those arguing that companies failing to move toward sustainable development will be at a competitive disadvantage.

Stakeholder theory is also pertinent to the design of an effective framework for strategy development (de Kluyver and Pearce 2006). Edwin Freeman originally defined stakeholders as "those who are affected by and/or can affect the achievement of the firm's objectives" (Freeman 1984, p. 137). More recently Jones, Felps, and Bigley stated that "stakeholder theorists view the corporation as a collection of internal and external groups [e.g., shareholders, employees, customers, suppliers, creditors, and neighboring communities]—that is, 'stakeholders.'" Relationships with stakeholders can be wrought with tension when organizational self-interest and stakeholder well-being are in conflict. Moral concerns can be a barrier to positive relationships with any or many of the stakeholder groups; the sense that the organization is dealing legitimately with stakeholders is considered by many stakeholder scholars as being "a fundamentally moral phenomenon" (Jones, Felps, and Bigley 2007, p. 141).

The consideration of the needs of stakeholders is in peril in many organizations. Phillips notes that "obligations of fairness" are created whenever parties are involved in a mutually beneficial arrangement (Phillips 1997). Unfortunately, the *authority principle* is alive and well in the twenty-first century. According to Thompson, people "are inclined to accept the opinions, directions, and admonitions of people we consider to be legitimate authority figures" (Thompson 1967, p. 36). The authority principle has been used to justify arrogating the "obligations" and privileges of making strategy to a few over the last century. There have been times when obligations of fairness were set aside and stakeholders were left at a disadvantage; the authority principle suggests that there be acceptance without objection. From our viewpoint, obligations of fairness are present in the application of an effective design framework for strategy development. Thus, the process used for designing sustainable development is significant to the outcome (Hutzschenreuter and Kleindienst 2006).

The following section offers a strategic approach for moving toward organizational sustainability development using the SOAR framework. This framework addresses many of the concerns of the leading strategic management scholars raised earlier in this chapter and builds upon some of the best strategic-planning practices.

The SOAR Framework

The SOAR framework for strategic inquiry and decision making was designed to bring a whole-systems approach into the strategic-planning process by making the organization's many stakeholders a part of the initial planning process and also a part of the ongoing activities involved in making the planning happen. Stavros, Cooperrider, and Kelley provide the language and framework to guide a strategic-planning process that focuses on inquiring into an organization's strengths and its opportunities to achieve high aspirations and measurable results (Stavros, Cooperrider, and Kelley 2003). Using SOAR in strategic planning helps organizations move from being mired in self-imposed constraints to discovering shared possibilities for moving toward becoming an organization that is committed to sustainable development. The SOAR framework in figure 2.1 goes beyond the SWOT model to link internal strengths and external opportunities to the vision and mission of the organization to create strategic initiatives, strategy, tactical plans, and measurable results. The four phases of the SOAR framework involve inquiring, imaging, innovating, and inspiring.

SOAR inquiry focuses directly on those elements that will give life energy to the organization's future and strategic moves. An organization's life energy and possible solutions are located in the conversations of its people. Life energy is made explicit in dialogue between and among groups of stakeholders (Holeman, Devane, and Cady 2006). The SOAR approach to strategy development and formulation starts with an *inquiry* into *strengths* and *opportunities*.

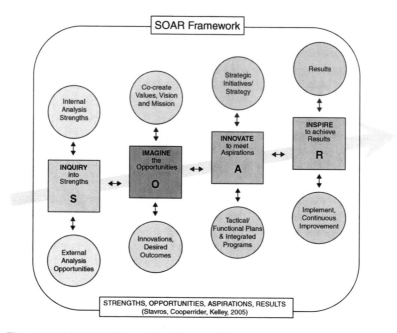

Figure 2.1 SOAR™: What we do and how we do it

Effective inquiry for strategic planning includes explicit consideration of the values and purpose of the organization—its vision, the customers it serves, and the stakeholders it impacts. The *inquiry* phase enables participants to examine, clarify, reflect, and consciously evolve their truly desired purpose and commitment to themselves and the organization.

The *imagine* phase is the one where strengths and opportunities are collectively explored and possible new ones emerge through stakeholder dialogue aimed at creating strategic alternatives. This phase also identifies the shared value set, vision, and mission of the organization through opportunities for stakeholders to be involved in the formulation of effective strategies (de Kluyver and Pearce 2006). The *innovation* phase is a call for *aspirations* that collective cocreate the most-preferred future. This phase also involves design of strategies and tactics. Finally, the *implementation* phase is the one where employees are inspired through authentic recognition and reward to achieve measurable *results*.

The SOAR framework is a search for the life-giving parts of the organization and for the elements that make those life-giving forces possible; it contains the *whole* of the organization. According to Rainey, "Exploring the life-giving forces of the firm should be made applicable. This leads to knowledge that can be validated and used productively" (Rainey 1996, p. 211). If CEOs want to embrace sustainable-development concepts in their companies' strategies and business models, the SOAR framework enables a wide variety of contributors to bring these concepts into the formulation and assessment of the organization's strategies. As new underlying business assumptions become apparent, such as ones calling for a sustainable world, those assumptions become more effective in supporting policies and practices consistent with sustainable development. One assumption may be that people want to operate in the best interests of people and the planet; they just need some assistance to overcome the old paradigms and they must devise new pathways to get there. The emergence of such an assumption contributed to and underlies the Center for BAWB's global inquiry into the business sectors that are putting their people and assets to work to benefit the environment (BAWB 2005).

Examples of Organizations Using SOAR

SOAR has been used successfully in the strategic-planning process in environmental scanning; in revisiting and recreating organizational values, vision, and mission; in formulating strategy, strategic plans, and tactical plans; and in bringing about transformational change. The framework has been used in for-profit and nonprofit settings and at the corporate and strategic business unit (SBU) level in education, manufacturing, service, health care, automotive, pharmaceutical, and banking. Roadway, Tendercare, Textron Fastening Systems, Positive Change Core (PCC), Utah Education Association (UEA), and CASE University have used SOAR in their strategic-planning efforts. Table 2.1 highlights organizations that have recently used SOAR for pursuing sustainability goals.

In addition to achieving measurable performance and business results, organizations using this approach have learned to

- integrate a whole-systems stakeholder perspective
- inquire into and leverage the strengths and opportunities of the organization in the best interests of its stakeholders
- create shared values, vision, and a mission
- develop a strategic and tactical plan that aligns with the organization's transformational forces
- identify the structures, systems, and processes to support the plan
- build dynamic relationships to implement the plans and sustain success

The most important part of any strategic-planning process relates to the question of sustainability of the designed path or course. Sustainability is

Table 2.1 SOAR sustainability applications

Selected Client Information	Brief Description of Work
Biological Conservation Charity	To create a strategic plan with priorities that serve as a living document.
BAE Systems	To create a division-wide strategic plan for an acquired division.
CASE University	To create a strategic plan and new brand identity for the university.
Cathedral Foundation	To bring stakeholders together to design a plan to serve all its community members.
DBC—National Healthcare Board	To use appreciative strategy processes in co-creating an engaging leadership team that develops a shared vision for its national planning board.
Fairmont Manufacturers	To discover sustainable manufacturing solutions.
FCI Automotive	To discover a strategy to improve supply chain management and inventory quality.
Hayes-Lemmerz–Cadillac	To discover how the plant can be environmentally profitable while decreasing operational costs and improving plant efficiencies while plant sales are flat.
Jefferson Wells	To engage the whole practice in strategy development and execution to create a positive impact on human metrics.
John Deere	To align strategy at corporate, business, and functional areas.
Metropolitan Library System	To bring about a strategic integration of a Chicago metropolitan library system.
Orbseal Plymouth Tech Center	To align a newly created technology center with corporate strategy. The parent division was physically relocated to a different state so it would be near its OEM customers.

Source: Created by Jackie Stavros from internal documents and stories on SOAR, 2000–Present.

best understood to have three components—confidence, momentum, and a balance of anticipation and responsiveness to a changing environment.

Confidence, Momentum, and Anticipation

Confidence, as noted in a book by Kanter, is the place between arrogance and despair, and it can be seen most explicitly in naturally occurring winning teams (Kanter 2004). According to Kanter, what characterizes winning teams is the feeling of *soaring*. But most importantly, confidence is a capacity for heightened positive perception of strengths. This is what the SOAR framework achieves. In the first phase of SOAR, the inquiry into the possible is all about the identification and mapping of strengths and magnification of opportunities. It is about the belief that the vision and tasks that are before the group can be realized. If sustainable development is the task, then this sort of confidence is essential to anticipation of the future and the ability to respond to the organization's environment on an ongoing basis.

Leach and Moon define momentum as the impetus gained by a moving object (Leach and Moon 2004). The *Business Line* website defines it in a business sense as "the fuel in the tank that drives strategy into action" (Murali 2004). A way to nurture momentum and sustainability is to set up a community of practice or a learning network that makes the search for strengths, innovations, and opportunities part of everyday thinking and acting.

A Fortune 100 financial management organization offers one example of a learning network focusing on the search for strengths, innovations, and opportunities. In one of the company's fifty offices, quarterly meetings were designed around the SOAR framework to inquire into the past three months' most powerful innovations and service deliveries. These meetings demonstrated the effectiveness of SOAR's strengths, opportunities, aspirations, and results in action. In the first year of using the SOAR framework, the organization's attrition rate fell from 24.5 percent to 12.9 percent. Today, it is 7.6 percent. The gains in human capital (HR metrics) brought the office from a strong position among the company's fifty offices to the top position two years later. As shown in table 2.2, the office's 2005 rankings were as high as third and as low as eighth on four metrics: two financial and two HR. In 2007 the office was ranked first on all four metrics. As the results suggest, this form of whole-system learning builds momentum and magnifies the confidence in its people.

Finally, sustainability development depends on being able to *anticipate the future and embrace* the changing environment. A leading aerospace company is using the SOAR framework to create strategic initiatives that can both anticipate and align with the uncontrollable political, economic, and environmental changes in the world. This company has learned that more important than sustaining the strategy is the need to nurture a culture

Table 2.2 Global professional service firm— three-year results

Financial Metrics	2005	2006	2007
Revenue	8th	5th	1st
Income	7th	2nd	1st
HR Metrics			
Retention	3rd	1st	1st
Engagement	6th	1st	1st

Source: Client Annual Report.

of *strategic learning and leadership where it can creatively balance anticipating future events while responding to today's events*. One of the benefits of repeated use of the SOAR framework is that organization members continue to seek strengths, seize opportunities, articulate aspirations, and assess results. These are the nutrients organizations need to feed the kind of learning that is relevant and adaptable.

Summary

In 2000, Monsanto's CEO, Bob Shapiro, was quoted as saying, "Business must urgently turn its huge talents to finding ways of operating that produce the goods and services—particularly food—we want, while doing no further damage to the environment, if business wants to stay in business" (Lewin and Regine 2000, p. 323). While sustainable development is still emerging as a major force in industry, the status quo is still deeply entrenched. The good news is that the status quo is a system of people; people who care about future generations. Each person can play a part in the system. Each person can change. Consumers drive the decisions of business.

Organizations must first have a clear understanding of the implications of sustainable development to their operations and long-term survival—and need to know that organizational and community members care about sustainable development. There also needs to be executive leadership in the organization that is willing to demonstrate a sincere passion to improving the environment. It will require a commitment to manage the envisioned change with an emphasis on combining new strategic frameworks such as SOAR that leverage a whole system approach. SOAR combines and leverages these approaches by using dialogue in a framework that builds upon an organization's positive core. By focusing on strengths and opportunities, organizations can reach their aspirations (desired outcomes) with measurable results by:

1. *Inquiring* into strength and opportunities;
2. *Imagining* the best pathway to sustainable growth;
3. *Innovating* to create the initiatives, strategies, structure, systems, cultures, and plans; and
4. *Inspiring* action-oriented activities that achieve results

This process involves creating a critical mass focused on attaining the values, vision, and mission of the sustainable (green) organization. It requires timely and effective training of organizational members to make this transformation a reality. It also requires the development of systems to ensure that managers are accountable for creating an environment of sustainable development. It is important to understand that this transformation requires a great deal of time, effort, and money and that short-term losses will likely occur as the process begins. For progress to continue, an ongoing dialogue on sustainable development must exist between organization stakeholders. This ongoing dialogue is of utmost importance. When done correctly, this dialogue will result in a sustainable advantage for the organization and an improved environment for all living creatures.

References

Amason, A. 1996. Distinguishing the effects of functional and dysfunctional conflict on strategic decision making: Resolving a paradox for top management teams. *Academy of Management Journal* 39 (1): 123–148.

Astley, W. G., and Van de Ven, A. H. 1983. Central perspectives and debates in organization theory. *Administrative Science Quarterly* 28 (2): 245–273.

BAWB: Business as an Agent of World Benefit. 2005. http://worldbenefit.cwru.edu/ (accessed January 7, 2007).

Boeker, W. 1997. Strategic change: The influence of managerial characteristics and organizational growth. *Academy of Management Journal* 40 (1): 152–170.

Burke, W., and Trhant, W. 2000. *Business climate shifts: Profiles of change makers.* Woburn, MA: PriceWaterhouseCoopers.

de Kluyver, C., and Pearce, J. 2006. *Strategy: A view from the top.* Upper Saddle River, NJ: Prentice Hall.

Dean, J. W., and Sharfman, M. 1996. Does decision process matter? A study of strategic decision-making effectiveness. *Academy of Management Journal* 39 (2): 368–396.

Dechant, K., and Altman, B. 1994. Environmental leadership: From compliance to competitive advantage. *Academy of Management Executive* 8 (3): 7–20.

Freeman, R. 1984. *Strategic management: A stakeholder approach.* Boston, MA: Pitman.

Frooman, J. 1999. Stakeholder influence strategies. *Academy of Management Review* 24 (2): 191–205.

Gladwin, T., Kennelly, J. and Krause, T. 1995. Shifting paradigms for sustainable development: Implications for management theory and research. *Academy of Management Review* 20 (4): 874–907.

Griffiths, A., and Petrick, J. A. 2001. Corporate architectures for sustainability. *International Journal of Operations & Production Management* 21 (12): 1573–1585.

Hamel, G., and Prahalad, C. K. 1996. Competing in the New Economy: Managing out of bounds. *Strategic Management Journal* 17 (3): 237–242.

Hart, S. L., and Milstein, M. B. 2003. Creating sustainable value. *Academy of Management Executive* 17 (2): 56–67.

Hawken, P. 1993. *Ecology of commerce.* New York: Harper Business.

Hitt, M. A., and Tyler, B. B. 1991. Strategic decision models: Integrating different perspectives. *Strategic Management Journal* 12 (5): 327–351.

Holeman, P., Devane, T., and Cady, S. 2006. *The change handbook*. San Francisco, CA: Berrett-Koehler.

Holliday, C. 2001. Sustainable growth, the DuPont way. *Harvard Business Review* 79 (8): 129–132.

Hutzschenreuter, T., and Kleindienst, I. 2006. Strategy-process research: What have we learned and what is still to be explored. *Journal of Management* 32 (5): 673–720.

Interface Sustainability. 2006. http://www.interfacesustainability.com/seven.html (accessed December 17, 2006).

Isaacson, W. 2006. Letter from the President. http://www.aspeninstitute.org/site/c. huLWJeMRKpH/b.493947/k.7AC3/About_the_Aspen_Institute.htm (accessed December 14, 2006).

Jones, T., Felps, W., and Bigley, G. 2007. Ethical theory and stakeholder-related decisions: The role of stakeholder culture. *Academy of Management Review* 32 (1): 137–155.

Kanter, R. 2004. *Confidence: How winning streaks and losing streaks begin and end*. New York: Random House.

Kassinis, G. I., and Vafeas, N. 2006. Stakeholder pressures and environmental performance. *Academy of Management Journal* 49 (1): 145–159.

Leach, J., and Moon, J. 2004. *Pitch perfect*. Hoboken, NJ: Wiley.

Lewin, R., and Regine, B. 2000. *The soul at work: Listen... Respond... Let go*. New York: Simon & Schuster.

McCarthy, D. J. 2000. View from the top: Henry Mintzberg on strategy and management. *Academy of Management Executive* 14 (3): 31–39.

Murali, D. 2004, September 16. Momentum is the fuel that drives strategy into action. *The Hindu Business Line (Internet edition)*, http://www.thehindubusinessline.com/catalyst/2004/09/16/stories/2004091600200200.htm (accessed February 19, 2008).

Nutt, P. C. 1999. Surprising but true: Half the decisions in organizations fail. *Academy of Management Executive* 13 (4): 75–90.

Phillips, R. A. 1997. Stakeholder theory and a principle of fairness. *Business Ethics Quarterly* 7 (1): 51–66.

Rainey, M. A. 1996. An appreciative inquiry into the factors of culture continuity during leadership transition: A case study of LeadShare, Canada. *OD Practitioner* 28 (1&2): 34–41.

Schweiger, D. M., Sandberg, W. R., and Ragan, J. W. 1986. Group approaches for improving strategic decision making: A comparative analysis of dialectical inquiry, devil's advocacy, and consensus approaches to strategic decision making. *Academy of Management Journal* 29 (1): 51–71.

Shrivastava, P. 1995. The role of corporations in achieving ecological sustainability. *Academy of Management Review* 20 (4): 936–960.

Stavros, J., Cooperrider, D., and Kelley, D. 2003. Strategic inquiry > Appreciative intent: Inspiration to SOAR. *AI Practitioner* (November), 2–19.

Suntop Media Thinkers 50. 2007. The Thinkers Top 50. http://www.thinkers50.com (accessed January 9, 2007).

Sustainable Development Commission. 2006. I will if you will: Towards sustainable development. http://www.sd-commission.org.uk/pages/publicationstop10.html (accessed January 2, 2007).

Sustainable South Bronx 2008. http://www.ssbx.org/greenway.html (accessed February 22, 2008).

Thomas, H., Pollock, T., and Gorman, P. 1999. Global strategic analyses: Frameworks and approaches. *Academy of Management Executive* 13 (1): 70–82.

Thompson, A., and Strickland, A. 2005. *Crafting and executing strategy.* New York City, NY: McGraw-Hill.

Thompson, J. 1967. *Organizations in action.* New York: McGraw-Hill.

Waddock, S. A., Bodwell, C. B., and Graves, S. 2002. Responsibility: The new organization imperative. *Academy of Management Executive* 16 (2): 132–148.

WBCSD: World Business Council for Sustainable Development. 2007. http://www.wbcsd.org (accessed January 9, 2007).

Wheelen, T., and Hunger, J. 2006. *Strategic management and organization policy.* 10th ed. Upper Saddle River, NJ: Pearson Education.

PART II

Innovations in Financial Thinking and Action:
Beyond the Primacy of Today's Stock Price

CHAPTER 3

Sustainable Wealth Creation beyond Shareholder Value

Mats A. Lundqvist and Karen Williams Middleton

Introduction

The university can be a place for generating returns on investments—returns that are both financial and of other nature. This chapter investigates an approach in which action-based masters-level education is integrated into university venture creation. The approach is resided at Chalmers School of Entrepreneurship (CSE) and Göteborg International Bioscience Business School (GIBBS) in Sweden. The purpose of the schools is to champion ideas into viable investment opportunities through a combination of venture development and entrepreneurial training. The schools promote the responsible utilization and commercialization of primarily university-based research ideas in the fields of technology and bioscience. They also package ideas stemming from individual inventors or from firms. This approach accommodates promising ideas and research results that are not so "low-hanging" as to be championed into start-ups. All this is done while shaping aspirant entrepreneurs during their masters-level education. Many of these aspirants will continue running the venture beyond its incorporation into a firm and often through several rounds of venture capital financing. This approach will be described and analyzed as an innovative way of accomplishing multiple returns on investments that contribute to sustainable development in several ways.

The education provided at these schools is intended to give the entrepreneurial student the opportunity to "test the water"—to go through real-life entrepreneurial and business activities in order to learn by doing, reflect upon the consequences of action, develop decision-making processes, and prioritize activities. The method for "testing the water" of start-up development is to actually allow students to become collaborating partners in and around ideas that have commercial potential, and then to guide and support the venture designed around the idea through the process of building a legitimate business. Students are encouraged and expected to actively seek out and test the skills and tools provided within the real-life context of the joint ventures. The ventures are constituted by the students while the idea provider(s) and the school representatives act as stakeholders, and all parts may be potential future shareholders. This experience of building a real venture empowers students to create value rather than just "earn income," as well as to become driven and motivated individuals or teams that can positively affect or influence society. At the school, entrepreneurship is not only about driving an idea toward economic success, it is also the prime tool for achieving economic, social, and environmental sustainability. This is part of a wider stream of sustainable entrepreneurship research and practice associated with sustainable entrepreneurship (e.g., Cohen and Winn 2007; Dean and McMullen 2007; Marsden and Smith 2005).

Tangible Results

The venture creation approach developed by CSE (for an overview of CSE, see *Innovating Regions in Europe* 2008) and adapted by GIBBS has generated tangible results in the following forms. Regarding formal education, more than 200 students have received masters-level degrees from CSE and GIBBS. More than half of the graduate students are now engaged in start-up companies. Nearly all of the remaining students engage in various forms of business development in established firms (often beginning as trainees), as business consultants, or within the so-called innovation system (i.e., as incubators, seed-financiers, etc.).

The so-called preincubator fund, operative since 2001, currently includes a portfolio of twenty-five graduated companies built from the idea-based ventures developed during the course of the education at CSE and GIBBS. The preincubator takes a 20 percent share in the potential start-up company. Within this share, the investors in the preincubator have a right to half of the capital generated, but not access to equity or ownership influence. The portfolio of companies had a market value of U.S.$80 million in the spring of 2007 and includes start-ups based in bio- and nanotechnologies, applied materials, medical diagnostics, and information and communication systems, among other things. In 2006, CSE and GIBBS portfolio companies attracted a collective U.S.$7.5 million in financing—primarily through equity investments. The 25 companies together employ more than

160 persons, with yearly turnover in excess of U.S.$20 million. The companies contribute to regional commerce through taxable revenue, collaboration with and use of regionally located partners and distributors, and volunteer contributions to their former educational institutes through lecture presentations and mentoring. A total of five companies founded at CSE and owned by the preincubator have made a successful exit from the portfolio. In recent years CSE and GIBBS have piloted ventures with companies such as Volvo, Saab, and StoraEnso and collaborated with researchers from universities in other cities, such as Stockholm and Oslo. Another measure of this success are the awards that graduated companies continue to receive; for example, CSE and GIBBS graduate companies have received one-third of the total twenty-four medals awarded through the Venture Cup West business plan competition thus far.

Going Beyond Traditional Shareholder Value

Developing schools that incorporate individual and venture development, with both educational and commercial ambitions, poses challenges. This chapter will focus on two questions that begin to address these challenges:

1. How do you secure educational objectives while also building ventures?
2. What returns on investments, other than financial, result from the approach?

To answer these questions, we first clearly present the educational structure and surrounding framework, giving concrete examples of how the structure affects the intention of the schools. Next, we give a short history of the evolution of the schools. Finally, we present illustrative cases showcasing the similarities and differences of some of the individuals from, and companies developed through, the schools in order to demonstrate how sustainable value can be generated. (An alternative approach, the Invention to Venture Entrepreneur Bootcamp held in Massachusetts, is discussed in Halpern and Plano [2006]).

The substantiating elements of the CSE and GIBBS approach can be organizationally understood in terms of the following:

1. a **masters-level program**—situated in an interdisciplinary and practical innovation system environment
2. a **preincubator**—a group of people who finance and manage the extraordinary efforts needed to recruit future entrepreneurs and develop innovation projects

3. a **venture team**—a group of key stakeholders
4. an **entrepreneurial network**—with alumni, with researchers, with innovation system and investment actors, etc.

These four elements are intertwined into the combined educational and venture creation approach. In part, the educational and preincubation contributions are depicted in figure 3.1.

Masters-Level Program

Providing the approach at the masters level is a conscious choice. At the masters level, students generally intend to pursue a commercial career upon graduation, whereas once engaged in doctoral studies, as the situation currently stands, individuals have often entrenched themselves into academic careers. Specially selected masters-level students also have sufficient education and legitimacy to generate interest around an early-stage venture.

The program emphasizes science-based entrepreneurship and business creation through real-life ventures, project assignments, IT-based simulations and role-plays, teamwork, and interplay with the university, its innovation system, and the surrounding knowledge-based industry. In short, the program offers the student a laboratory for simulated and real-life action learning (course structures are depicted in figure 3.1). This laboratory was built around CSE since 1997, with contributions from the Center for Intellectual Property studies (CIP) since 2000 and with the addition of GIBBS in 2005. Both CIP and GIBBS are joint ventures with Göteborg University, specifically with its business school and medical school, emphasizing development toward more interdisciplinary environments.

CSE and GIBBS stem from high-technology ideas/inventions, with GIBBS specializing in bioscience. CSE and GIBBS students are expected to have a high level of motivation for and interest in technology-based innovation projects, including interaction with idea providers (inventors and researchers), fellow students, and international experts. Students with backgrounds in engineering, science, business, and law attend the program. The balance between engineering and science on one side, and business and law on the other, is approximately 50-50, if you include industrial engineering students in the latter category. This diversity provides an environment with opportunities for students to constructively learn from one another while enhancing their specific strengths in innovation and venture creation.

Since its start, CSE (and later GIBBS) has continuously developed its specialized student recruitment process. Student applicants go through three stages of selection: a review of base criteria, a written application that includes

New Venture Strategy and Formation, 7.5 credits

Entrepreneurial Leadership and Organizing, 7.5 credits

Valuation and Entrepreneurial Finance 7.5, credits

Marketing a Knowledge-based Business, 7.5 credits

Innovation Project and Master's Thesis
30 credits

Idea Evaluation and Feasibility Study
7.5 credits

Elective Courses

Elective Courses

Elective Courses

Technology-/ Biotechnology-based Entrepreneurship
15 credits

Intellectual Property Strategies
7.5 credits

Design of Technological Innovations and Markets
7.5 credits

| Q1 Year 1 Q2 | Q3 Year 1 Q4 | Q1 Year 2 Q2 | Q3 Year 2 Q4 | Q1 Year 3 Q2 |

├ PROJECT PERIOD ─────────────────── COMPANY FORMATION ─┤

CSE INCUBATION PROVIDES:

- Project recruitment, evaluation, and selection by a Project Committee. Final selection together with the CSE and GIBBS students

- Structure, process experience, and business development know-how
- Start capital SEK 25,000 + possibility to raise additional SEK 75,000
- Office facilities and equipment

- Industry professional as chairman of the project
- Business coach from CIP Professional Services
- Alumni coach for each project
- Network of Business Affiliates

- CSE and GIBBS students often continues as business developers and are the primary driving force
- Ownership by CSE Holding AB, idea provider, and potentially students
- CSE Holding AB member of the board

Figure 3.1 Program structure

essay questions, and interviews. Base criteria ensure that the applicant has fulfilled the required undergraduate education within accepted areas of education and shows sufficient English and computer skills. The accepted areas of education are relatively broad but have some restrictions because of the fact that the venture projects are based on high-technology research ideas. Essay questions, together with other supplementary information, including a letter of recommendation and CV, are used to test the applicant's ability to communicate ideas, accomplishments, decisions, and experiences in a reflective and structured way. Questions used have been benchmarked with other essay questions used by comparable entrepreneurial educations and are designed by behavioral scientists/psychologists to provide a forum in which the applicant can communicate the following traits:

- Motivation and commitment to the unique action-based learning process
- Personal responsibility and awareness
- Ability to handle risk and complexity
- Leadership
- Ambition
- Effective communication skills

In addition, a student should also be characterized in at least one of the following roles:

- Visionary
- Team builder
- Efficient user of resources
- Analyst

Student applicants who fulfill the initial base requirements and effectively communicate a majority of the characteristics and skills listed above are then required to attend interviews with selection committee representatives. Applicants are interviewed both individually and in case format (to observe their reaction to team dynamics and their individual identity within a team). Individual interviews consist of questions to ascertain information regarding motivation, experience, leadership, teamwork, risk and uncertainty, creativity, independence and responsibility, and decision-making processes. At the same time, the interviews are a forum for student applicants to pose any questions they may have regarding the structure and format of the program as well as an opportunity for interviewers to communicate expectations and requirements of the program. In the final selection meeting all of the above aspects are weighed before a decision of admittance is made. Notes are made with a synopsis of the reasoning behind each decision.

Preincubator

The preincubator has operated since 2001 and consists of two fully owned daughter companies of Chalmers University of Technology. This construction avoids conflicts of interests in order to minimize risk, particularly on the individual level, especially during the innovation project year period. The main duties of the preincubator are to manage the recruitment and development of ventures that could be future companies for CSE and GIBBS, to provide initial seed financing to accepted ventures with secondary financing should the ventures prove qualified, and to eventually own stakes in the companies started from the ventures. This preincubator is fully integrated operatively with the educational program. The preincubator also facilitates long-term development of the ventures and companies, additional procurement of resources, support of alumni activities, business development opportunities, marketing and outreach programs, and other activities that benefit the students, companies, and schools.

To produce start-ups, the preincubator screens more than one hundred ideas every year. Of these, more than ten per year are selected to run as an innovation project in the combined preincubator and education structure, with additional ideas held in reserve should one or more of the ventures be terminated. Figure 3.2 illustrates the recruitment and idea-flow process for CSE 2005, during which projects were terminated and replaced, thus allowing for a total of thirteen ideas to be thoroughly pursued as ventures and eventually five companies to be founded at the end of the educational period.

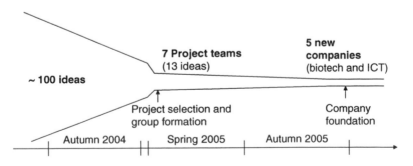

Figure 3.2 CSE 2005 idea recruitment deal flow

Venture Team

Five types of stakeholders constitute the CSE and GIBBS venture team: students, educators, idea providers, representatives of the preincubator, and the venture board chairperson. The students constituting the venture team are, of course, key stakeholders and drivers in the learning and venture creation process. They are both guided and empowered by the other stakeholders, while partnering with these same stakeholders to pursue entrepreneurial processes and potentially to build the venture into a company.

Educators come in two main forms: those directly linked to the school, working at the school on a day-to-day basis with a cognitive understanding of the complexity in which the students act, and those that are partners to the school, operating in other educational departments or in industry, and who are thus capable of providing experience- or academic-based knowledge in specific areas. Among the direct educators three have three or more years' experience in start-ups, three have ten or more years' experience in organizational development and leadership, three have four or more years' of legal experience, and two have ten or more years' experience in strategy and marketing.

To fulfill the ambition of creating high-tech companies, the students are formed into teams and are linked with an idea provider, the third key stakeholder in the school. Idea providers are contractually conjoined to the school on a case-by-case basis, ensuring both their participation in the venture and student development while also protecting their interests in the ideas they initiate. Idea providers often provide deep technical insight and often co-supervise technical studies in the students' theses, thereby being key bridges in integrating science, technology, and business.

Supporting this structure is the fourth stakeholder in the process, the different representatives of the preincubator. These individuals are active on the management board of the venture during the educational program with the purpose of supporting the best interests of the venture and upholding the perspective of the preincubator. Each venture establishes a board, including the fund representative and idea provider, and selects a chairperson of the board, the fifth key stakeholder. The chairperson is chosen on the basis of industry expertise, as it relates to the project, business development experience, and program interest (i.e., the individual is willing to allocate time and energy to the management processes because he or she is associated with an educational process).

Entrepreneurial Network

Extenuating from these key stakeholders is then a network of other actors, with various degrees of connectivity to the schools, for example, business angels, international advisors, mentors, and other incubation actors. This group provides information and support, through which the progress of the students and the potential companies is accessed and advanced. Currently CSE and GIBBS students, apart from tapping into an extensive network, also receive systematized coaching from an alumni coach, advice in intellectual asset and property management from collaborative consultants, and legal as well as accounting advice from Göteborg accounting firms and law firms that offer pro bono time to the students, with expectations of gaining them as customers in the future.

Historical Background

The inspiration for CSE grew from the assumption that a great number of good ideas fail to become business ventures and thus are lost to society. Of all the components needed to start a new venture—including good business ideas and venture capital—start-up entrepreneurs were assumed to be the greatest scarcity. Researchers and academics are found to rarely champion their own ideas to the market, even though they are entitled to them through the so-called teachers exemption, giving ownership of research results to professors, if not otherwise agreed upon. CSE identified its position in the gap between invention and a viable investment opportunity (see figure 3.3). Within this gap, selected entrepreneurial students and selected ideas are brought together through the approach previously described.

In the late fall of 1995, Mats Lundqvist and Sören Sjölander, of the department of technology management and economics at Chalmers, decided to create a school that would arrange for partnerships between inventors with ideas and students with the drive to become start-up entrepreneurs. It became apparent that most existing entrepreneurship programs

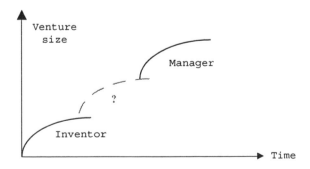

Figure 3.3 Reason for starting Chalmers School of Entrepreneurship in 1997

were focused on teaching entrepreneurship rather than on actually developing entrepreneurs. Since its start in 1997, CSE has had the double mission of developing entrepreneurs while simultaneously creating technology ventures. At Chalmers, a good breeding ground for such an organizational innovation existed thanks to a constructive engineering culture combined with a high appreciation of research commercialization, pioneered by Torkel Wallmark and others as early as the 1960s.

The program at CSE can be seen as a continuous development that has involved hundreds of committed contributors. Learning from each year of activities has resulted in three main stages of core development, distinguished as follows:

- Version 1 (1997–2000), in which special project and student recruitment processes were developed along with a project-based pedagogy located in Chalmers Innovation incubator environment. This version was a final-year program for engineering students at Chalmers.
- Version 2 (2001–2004), in which CSE became a one-and-a-half year masters-level program recruiting broadly from all over Sweden, and a special holding and incubation company was started together with AB Chalmersinvest.
- Version 3 (2005–2007), includes the starting of the sister school, GIBBS, and a first semester integration, called Business Design.

Within each version, certain challenges have caused changes in the approach, as have accomplishments that reinforced the design. In the following section, specific cases of individuals and companies that have graduated from CSE and GIBBS are presented to illustrate the experience and development provided through the approach. Examples presented in the cases will enable discussion around the two questions posed earlier in the chapter.

Cases from CSE and GIBBS

Anna

In 1997, Anna Weiner responded to an invitation to apply to the first class of what was to become CSE. She and eleven other students, with various backgrounds from Chalmers University of Technology, embarked on a yearlong journey of entrepreneurial learning. Teamed up with two other students, Anna took the lead position in the venture idea provided. Throughout the year, Anna faced challenges in gaining authority regarding the venture.

At the end of the program, the venture was incorporated with Anna acting as CEO.

Still an inspired entrepreneur in 1999, Anna teamed up again, this time with a biochemist who had developed a specialized probiotic gel, to help found the company ELLEN AB. At the time the company was founded, initial clinical trials had been conducted and a patent filed. Additional trials were conducted in 2001, and the first products were launched one year later. ELLEN expanded to other Nordic countries in 2004, and internationally one year later. Once its CEO, Anna now works part-time at ELLEN; her main responsibilities are in research and development and strategic business development. Apart from her duties at ELLEN, she also advises multinational corporations on intellectual property, management, and financially related issues. The story of ELLEN, and Anna, is commonly used in the CSE and GIBBS courses as an example of the birth and growth of an entrepreneurial company with its beginnings in university technological research that carved out market space from well-established, multinational brands through the provision of a biologically conscious product.

Vasasensor

Vasasensor AB came to CSE in 2003 as a spin-off from the Imego (Institute of Microelectronics in Gothenburg) research institute. The student team was presented with a sensor patent application, which had several possible for applications. The students were given the challenge to find and develop the best possible application and market space within one year at CSE. There were multiple potential markets that the team could readily identify, but they were searching for the application with the largest market potential within a short period of time, building on the competencies of the team. After only a few months, the decision was taken in mid-March 2003 to aim the company toward the paper manufacturing industry and its requirements. The core team presented the decision to the board and never looked back. By the end of the year, Vasasensor AB, together with actors in the paper industry, had focused on developing a wireless sensor system for process optimization in paper mills. Their success has drawn interest in the technology from several other industries as well.

The company continues to work toward bringing innovative technology to a more traditional industry (paper manufacturing), saving money and also time and energy sources. Vasasensor AB continues to work with local factories to develop additional prototypes, which may eventually be licensed to other parts of the world (but allows for the know-how to be developed within western Sweden). At the same time, the management team travels to other parts of the world (e.g., to Asia last fall), spreading the story of a company that was started through CSE. Sofia Johnsson, CEO of Vasasensor AB, and fellow CSE classmate, teammate, and Vasasensor management member, Brodde Wetter, regularly contribute to the further development of CSE and GIBBS, both at the masters level and in continuing educational programs developed for entrepreneurial alumni.

Denator

Most often, the ideas from which the ventures are based come from university researchers, independent of the students. However, the idea behind Denator AB, stemmed from research conducted in part by a family member of one of the CSE 2005 students. The family dynamic helped to fuel the growth of Denator during the educational year of the venture, with the team and support system striving to meet and champion developmental needs. Denator utilizes a patent-pending physical method to completely stop the degradation process in biological samples that enables a clearer picture of the proteome. The company created a chain of products that ensure quality preservation of biological samples for protein analysis. The company also developed a biomarker for sample quality on the protein level, which is now being developed as the first sample quality control method in the proteomics field. The products are designed to enhance proteomic research in academic and pharmaceutical research, with further aims to implement the products at hospitals in order to improve diagnostics.

Recognizing the challenging and time-consuming time to market of bio-based ventures, attracting financing has been critical. As of January 2007, Denator has managed to receive approximately SEK 1 million in grants and competition financing, as well as two rounds of capital financing—SEK 5 million in the first round from one company and SEK 15 million in the second round from a mixture of private and venture-capital resources. Capital is utilized to expand operations, including the opening up of an additional office in Uppsala (where the idea originated), as well as on fueling production development.

Denator is utilized as an introductory case in the first semester of the program to present the complexity and challenges of a biotechnology start-up to new students; the management team visits to speak about the journey and the steps at the end of the case study. The core entrepreneur experienced several challenges during his year at CSE, resulting in development of additional coaching support being implemented into the education. At the same time, the entrepreneur has continued to seek out coaching and advice from some of the educational staff, a practice that perpetuates mutual learning and facilitates additional understanding of the entrepreneurial needs of the students into the educational pedagogy.

Ecoera

Based on an industrial idea but developed through collaboration with university-research arenas, Ecoera provides solutions for second-generation biomass heating fuel: agropellets. The current problems associated with gases released during the combustion of residual biomass are minimized through the use of specialized pellet formulations. The reduction of undesirable by-products combined with the use of an alternative and abundant resource stands to help standardize agropellet formulations and enables a new era of biomass utilization.

Ecoera AB was a latecomer to the CSE environment in 2006, starting as a new venture for a team that had terminated its previous project. The student team had only four months of official incubation (compared with the common one-year incubation period). In fact, the team was able to remain in the educational incubator space until the following spring, though the full network support provided by the educational platform was diminished in order to focus attention on the next generation of ventures. The extension past the "graduation" date also facilitated a management transition in the project team, where one of the original CSE teammates left the project and another graduate from the same year, but from the GIBBS program, took the vacant position. With limited incubation time available, the team was in critical need of research validation and the resources to carry out such validation. Network connections at Chalmers enabled adaptation of the project into another EU-research-funded project, which could be utilized as a unique early-stage financing method that encouraged the technical development of the agropellet.

Termination or Restart

While company formation does not occur in every case, the intention is to always provide value at some level. There are essentially three alternatives to incorporation of a venture into a company: noncommercial development, restart, and termination.

With early-stage ideas and ventures, analysis often shows immediate residual commercial value. In some cases, the venture is not currently adaptable to a commercial end but can be structured instead as a research or information platform. One project at GIBBS—the Chemical-Biology platform—is an example of such a venture (project). The ChemBio-project basically aims at creating a laboratory for medical research, focusing on high throughput screening of new substances. Although such a research platform might eventually have revenue-building intellectual property, its main benefit is in its enabling researchers, companies, and others to increase the efficiency of their research and innovation processes.

In other cases, the venture appears to have commercial value, but it is still too early in development to legitimize incorporation. Thus an idea-based venture might "walk over" from one team and project-year to the following year. The framework of the schools allows this to be possible, because no one—not even the first student team—lays any claim to the venture and the idea provider most often wants to extend the collaboration.

Even in the cases where a venture is going to be terminated, the venture can still provide residual value. Student teams are required to perform proper due diligence when providing reasons for termination, thus facilitating additional learning from the expected lack of potential of the venture—for example, the venture may be technically but not financially viable. Effective communication of the reasoning behind termination of the venture may allow an idea to be returned to the researcher in a way that facilitates further development and leaves open the possibility of commercialization at a later stage. Such due diligence also builds trustworthy relationships with idea providers, enabling further contacts for additional ideas or advice toward other ventures.

Securing Educational Objectives and Investigating Different Benefits

Earlier in this chapter, we presented substantiating elements of the approach represented by CSE and GIBBS, as well as examples of the outcomes, continued development, and evolution of the approach. The approach recognized the value of getting a financial return on investments in venture creation and goes beyond traditional shareholder value toward sustainable creation of wealth—economic, social, and environmental. We will first discuss how educational values are secured during venture creation (Question 1). Then we will analyze return on investments in four main categories (Question 2): financial, development of an entrepreneurial ecosystem, characteristics of companies created, and the professionalization of entrepreneurs educated to uphold sustainable development.

Question 1: How do you secure educational objectives while building ventures?

Securing educational objectives while building ventures requires insight into both the university arena and the business development arena. The founders of the schools understood that there was a gap between idea development and the formation of these ideas into business models or companies that would be accepted by the marketplace. This gap required the simultaneous development of individuals and companies so that the learning was intrinsically linked with the idea of development. However, the development process needs to be controlled and regulated to ensure that learning takes place. The founders realized they needed to establish school policies for the types of business ideas that would facilitate learning and for the structures and practice that would ensure learning was protected and championed during the development process.

The first step to ensuring educational objectives is to communicate the learning process as early as possible. Hence the philosophy is presented to potential students even before they are admitted to the schools through the specially designed application process. Through the written essay questions, potential students can communicate their intent and interest in entrepreneurial development. Interviews provide students with the opportunity to verbally communicate their intentions and for interviewers to discuss and steer expectations about schools, their format, and education prior to engagement. The essays and interviews help create a rapport between students and school staff which helps them find flexible solutions to conflicts between education and venture creation. This relationship is nurtured during the introduction to the schools when developing the social contract between the students and other stakeholders is emphasized. The social contract process is repeated at each major milestone within the education.

The general learning outcomes that students are expected to achieve on having completed CSE or GIBBS are that they are able to

1. **construct knowledge-based business** in interplay with complementary competences, thereby integrating technological, economic, managerial, and legal skills into innovations, products, ventures, and market offerings;
2. **analyze, construct, and use tools to design innovations**, such as different intellectual property tools (patents, standards, contracts, designs, trademarks, databases, copyrights, etc.), in interaction with research, market assessment, and product development;
3. **communicate, reflect, and manage group dynamics and responsible leadership** as applied to real-life and simulated complex situations;
4. Consider **citizenship and entrepreneurship for sustainable development; and**
5. **create** and manage **start-up** ventures.

In the first half year, theory is mixed with simulation exercises, under the requirement that the students build a basic entrepreneurial skill set prior to the action-based learning within the ventures and the venture teams. This first half-year mirrors more traditional and generally accepted methodologies of teaching. The schools have attempted to pioneer ways in which to examine learning outcomes that are not so easy to assess with traditional pedagogy, such as written exams. The examination of learning outcomes related to the application of skills and "reflection in action" has turned out to align with recent developments in European higher education, called the Bologna process. In this process, not only is higher mobility achieved by conforming bachelors and masters degrees to a three-year or two-year format, respectively, but learning outcomes that emphasize more vocational skills are also emphasized, hopefully contributing to employability.

During the innovation-project year, several mechanisms for school-level learning are utilized at CSE and GIBBS. Each of the mechanisms is intended to link venture learning to classroom learning, supporting the following objectives set out by the school.

1. **Role plays** in which the students act and "negotiate" with their classmates, with alumni, or with invited business people, provide a strong learning mechanism for CSE and GIBBS, particularly in relation to the activities of the venture creation process. Role plays allow for practice and for reflection upon actions carried out when acting in the venture.

2. Two **Project Reviews** and two **Business Reviews** are conducted during the innovation-project year. These arenas for presenting venture applications, business models, market segments, plans of action, and company value function as a "convergence point" for the holistic venture-based learning. The Project Reviews are internal, in that documents, business plans, and discussions are assumed to be more open and problems can be discussed. The Business Reviews are more open arenas where students are expected to be able to communicate and convince people such as future investors, of the values of their ventures.

3. Individual and team-based **assignments** are often applied on or inspired through the venture activities. A great deal of the rhythm of the education is driven by assignments, which can often bridge the gap between courses and, because of the nature of the real-life learning provided through the ventures, between theory and practice.

4. In the **school project** the whole school class organizes, finances, and executes a joint project, often including outreach to industry and secondary school education as well as school marketing, study tours, and other activities. In 2007, the CSE school project included a trip

to Uganda on which the class together with the Red Cross helped start a solar-panel-driven incubator. The GIBBS school project involved touring bioscience and biotechnology business, incubator, and university programs in California. The classwide project, like the open-office environment, provides for cross-venture learning. At the same time, the project allows the students to develop as entrepreneurs, by talking about their ventures, acting as ambassadors for the ventures and the school as a whole, and expanding their network.

Most instructors draw upon the above activities to demonstrate certain learning outcomes. In addition, each course adds its own measures, such as exams, course-specific assignments, and presentations. Another aspect of securing educational quality is the large amount of coordination done at the program level and not only at the course level. This coordination is partly IT-based through the adaptation of an open-source software facilitation platform, in which students and educators can deliver assignments and feedback; write journals; pose questions, answers, clarifications, and administrative information; present grading; and other learning facilitation activities. In addition to these, learning is facilitated through the following non-graded mechanisms, which still provide critical links between learning objectives and venture activities.

1. The **open office preincubator environment,** provided by CSE Incubation, enables cross-venture and cross-school (CSE and GIBBS) learning, as well as offering all the tools (phone, Internet, fax, meeting rooms, etc.) necessary to drive a start-up. Often the original venture teams of three students multiply in this environment with theses students and other potential key persons joining the venture during the innovation project year.

2. In three personal and three team-based **development talks** group dynamics, venture dynamics, learning, well-being, and other challenges are put forward by either the students or the educator. These talks are not graded in order to provide a safe and open forum for the students (the level of openness determined by the students and student teams themselves) to discuss and deal with issues they may face.

3. **Board meetings** are held in which school and preincubator staff document or reflect on how students prepare and execute the meetings, while students test their ability to communicate strategic direction of the venture to a board.

4. **Alumni activities** and "inter-year" learning. CSE and GIBBS have an active alumni association—Elumni. With regional development funding

beginning in 2005, CSE has leveraged the spontaneous networking among alumni by providing alumni leadership courses and linking alumni into the current program. As a result, CSE—alumni and later GIBBS alumni—are provided with continuing and strengthening support to help them develop a sustainable and appreciative leadership. Developments discovered through the course of the alumni's education are implemented back into the schools, in part through alumni acting as mentors for current venture teams.

Specific elements that secure learning objectives, should they conflict with business interests, are:

1. **venture-team formation** conducted by educational and preincubation staff to facilitate complementary and supportive skills within the venture team;
2. **unilateral right to venture termination** by the school, should the learning objective of using the venture become counterproductive to learning (such as the venture idea being determined as not having commercial potential); and
3. **investment restrictions** so that investment is limited to nonequity investments until the education is complete, to prevent, for example, conflict of interest.

A sophisticated team-formation process matches students to teams, and teams with a venture idea provided by the preincubator and selected by the class as a whole. The class selects projects through a process established independently by the class. Then the students individually list and rank ideas and competencies they would prefer to work with during the innovation-project year. This process balances the students' need for self-commitment with the construction of teams with complementary competencies, while utilizing all of the venture ideas selected by the class as a whole in the formation of the venture teams. It is within the team construct, and the different and often complementary perspectives of the team members, that much of the personal developmental learning takes place.

A three-party contractual agreement between key shareholders ensures their engagement and continued contribution to development (such as board meeting participation), and guarantees that venture activities will not fundamentally undermine educational objectives. The students, researcher(s), and preincubator all own shares in the venture, with the shares allocated to the chair through the preincubator should the venture be incorporated into a company after graduating from CSE/GIBBS. No single party owns a share greater than 49 percent. This clearly and contractually agreed-upon division of shares in the initial collaboration contract helps secure that only limited time is spent on negotiations between the future shareholders during the educational period. Many university incubators also appreciate the degree of structure and transparency the CSE and GIBBS preincubation process has in its current form and see it as a role model even from a strict business-creation perspective. The contractual agreement also requires a specific minimal-time engagement between the idea provider and the venture/team that ensures critical knowledge transfer is provided to the venture, so the idea can indeed be explored and developed to the full extent possible within the school period. Finally, the agreement clearly communicates that the school has the unilateral right to terminate the venture, should there be circumstances that are counter to the learning objectives. Taken together, these policies and measures have, throughout the years, proven to be critical in securing educational quality.

Question 2: *What returns, other than financial, on investments result from the approach?*

Because educational development and venture development are linked, the returns on investments expand beyond the direct financial gains to include other returns, such as building an entrepreneurially minded culture and a network, willing to give back in kind to the schools. These indirect returns contribute substantially to the development of CSE and GIBBS and to their larger university and business environments, that is, their entrepreneurial ecosystem. Beyond this, the ventures at CSE and GIBBS yield societal returns by commercializing innovations that contribute to sustainable development but that are too "high-hanging" for market and other actors to realize. At the very least, the experiences and competences gained by CSE and GIBBS students will provide returns to society beyond the potential formation of a company. The life-long entrepreneurial careers that CSE and GIBBS students pursue will arguably result in critical returns to society in the sense that they have developed the confidence and mindset to drive change and innovation for sustainable development. The launching of professional entrepreneurial careers, instead of traditional careers as employees in established structures, can be expected to have an important impact on wealth and welfare creation in knowledge-based economies.

The approach described in this chapter has built upon expectations and mechanisms for traditional financial returns on investments. By taking a share in every venture incorporated into a company, CSE and GIBBS control a slice of the financial value creation of the venture. Idea providers and students, having invested "sweat equity," will also become shareholders in the company formed. In addition, other individuals identified as key actors for the venture during its incubation at CSE and GIBBS will be compensated with shares of the newly formed company.

CSE and GIBBS have chosen to take equity in the ventures incorporated into companies for the following reasons:

- **To secure the operation and development of CSE and GIBBS.** In the short term, this is done by utilizing the investment money coming in from regional actors in the preincubator. In the long term, some of the companies formed will hopefully have a high rate of return when making an exit, thereby bringing back to CSE and GIBBS, and its investors, substantial financial return on investment.
- **To build relationships and learning with the financial community.** By being an active owner of the company portfolio, CSE and GIBBS are forced to be constantly responsive to the demands of the financial market, especially that of the venture-capital market. While CSE and GIBBS have strategically chosen to avoid having private investors in its core business, the schools—as stakeholders—still need to monitor their investments and ensure continuous, financial attractiveness until the companies become cash-flow positive. Linking up to the financial community legitimates that CSE and GIBBS are living up to their missions to create viable investment opportunities.

The building of entrepreneurial ecosystem is a critical indirect return on investment into CSE and GIBBS. School staff not only focus on the dynamic internal development of the functional aspects of the school but also engage in multiple external arenas of entrepreneurial development. They do this not only at the university, the regional, and the national levels, but also at the international level through research, by sharing best practices, and through developmental projects. These activities, conducted in collaboration with the students and the CSE and GIBBS networks, enable the ecosystem to continually evolve. A critical portion of CSE and GIBBS staff's time is spent on networking and coordination activities. Throughout the years of developing CSE and GIBBS, the return on such investments can be measured in several ways:

- Concrete partaking of the entrepreneurial network actors in the coaching of new students and ventures
- Alumni and other network actors returning to the schools as idea providers
- An entrepreneurial culture development from one year's class to the next, partially measured in terms of how attractive the ventures are for

investment and what is accomplished marketwise, technologywise, and otherwise in the ventures during the program.

During the ten years of operations, persons associated with CSE and GIBBS have taken on roles in virtually every part of the emerging "innovation system" of Gothenburg. Concretely, this means that the schools have been more or less crucial, contributing to and supporting initiatives such as business-plan competitions, business-angel networks, incubators, and seed and venture capital. Apart from these structures being important for CSE and GIBBS and their ventures, the structures are critical for the overall innovativeness of the region. In 2005, Gothenburg was recognized as the most innovative region in Sweden by the major Swedish technical newspaper *NyTeknik*. This was partly because of the developments at and around CSE and GIBBS. It was also due to large R&D intensive firms such as Astra Zeneca, Ericsson, and Volvo. These large multinationals still contribute in a more substantial way to the economic development and welfare of the region than do the companies from CSE and GIBBS. However, the increased ability of politicians, journalists, regional developers, and others to appreciate incremental as well as more radical innovations, is a notable aspect of the region today. In less than ten years, regional development authorities have diversified their investments to include company formation and entrepreneurial developments.

The examples and cases mentioned in this chapter provide evidence of how the entrepreneurial ecosystems functions. Alumni contribute directly to the education as lecturers, coaches, board members, and so on. Their companies are cases for new students to learn from, such as the Denator case presented here. Alumni, when seen in news and media, also contribute to the sustainability of entrepreneurial culture and ecosystem development as role models and as proof of concept for not only CSE and GIBBS students and CSE/GIBBS investors, but hopefully for others as well.

Beyond the local effects of the entrepreneurial ecosystems, CSE and GIBBS companies contribute to sustainable development by commercializing promising but "high-hanging" ideas. This is a societal return on investment, in that not only research results but also promising ideas from multiple segments of society are eventually brought to the marketplace (commercial or otherwise). As awareness in society increases in regard to the adoption of more sustainable technologies, CSE and GIBBS are an additional mechanism enabling such technologies to become more viable.

The case of Ecoera (http://www.ecoera.se/), which deals with agro biofuel, constitutes one example of how the barrier of abandoning fossil fuels in favor of more environmentally friendly alternatives can be lowered. In this case, CSE has the ability to develop something that the idea-providing company has no resources to do itself. CSE also connects the idea with academic research at Chalmers Department of Chemistry, which in turn could help prove and legitimize the sustainable technology.

Many other ventures at CSE, such as the above-mentioned Vasasensor, contribute to sustainable development in more indirect ways. As described in the Vasasensor case study, the wireless sensor technology allows the paper-pulp companies to increase their economizing in energy-consuming processes. In analogous ways, CSE ventures Vehco and Avinode help economize truck and business-jet transport using Information and Communication Technologies (ICT) technologies. Vehco is Sweden's current market leader in truck telematics, allowing truck drivers and truck owners to communicate, measure, and improve fuel consumption, among other things. Avinode is the leader on the European market for brokering and optimizing business jets for small and medium-sized companies.

CSE and GIBBS ventures constitute clear examples of how investments in education and venture creation can generate sustainable development. Arguably, the most important returns on these investments are the careers that the alumni pursue. Although it may still be too early to judge, indications are that alumni from CSE and GIBBS take responsibilities beyond running a single start-up. The indications include the following observations:

- Some alumni, such as Anna, have pursued several ventures as presented in the cases.
- Alumni are trained to argue for the public-good aspects of their ventures. This is partially done in order to attract soft loans or research money, as in the Ecoera example.
- The "school project" at CSE and GIBBS includes aspects of outreach and citizenship that over the years have become a central part of the school identity.

These indications build the argument that the approach described in this chapter has substantial and long-lasting effects on the professional identity of those graduating from the school. This combined with sustained collaborations and activities with the alumni assures that an entrepreneurial-professional identity concerned with innovation, change, and sustainable development will prevail.

Sustainable Wealth Creation

This chapter has investigated a university-based approach that combines education and venture creation around promising ideas. The approach has been developed over more than ten years. Tangible results as well as the daily operating of the two schools—CSE and GIBBS—have been accounted for. Two questions have been analyzed: (1) How do the institutes secure educational qualities while dealing with real venture creation? And (2) what different returns does the approach offer to shareholders in the specific ventures and beyond?

While venture creation and creating viable financial investment opportunities are at the core of the approach, it is not only financial returns to the shareholder that are relevant. Returns of a different character have been obtained thanks to the careful integration of educational and venture-creation activities. The approach has been critical to building an entrepreneurial ecosystem around the schools, which affects the academic as well as business environment in the region and beyond. Today this ecosystem is also critical for the development and running of CSE and GIBBS. Instead of education producing students who engage with society in a linear way, CSE and GIBBS are in constant mutual exchange with society both businesswise and learningwise. Unlike traditional education that often collects evidence from the real world and reproduces it through cases and theories in the classroom, CSE and GIBBS, in collaboration with their partners, constitute and create real-life cases that generate both value and learning (cf. Pretorius, Nieman and van Vuuren 2005).

Beyond the financial returns and the returns from mutual exchange within the entrepreneurial ecosystem, this approach also produces sustainable development on two levels. First, the ventures themselves are built on ideas that generate sustainable development, either directly, through commercializing new and more environmentally friendly technologies, or indirectly, by helping established technologies and processes become more efficient and monitored. Second, and perhaps most importantly, the graduates from CSE and GIBBS will likely continue to contribute to sustainable development through innovation well beyond their first or second venture, and as role models to others.

Ten years of experimentation has provided substantial learning and evidence about the benefits of the approach. Nevertheless, it is still in its early stages. Although substantial energy and time have been spent spreading the approach outside the university, traditions of teaching, research, and university management are still far from changed. Today, many entrepreneurship programs are applying an action-based pedagogy, more or less linked to the technology-transfer activities of the university. The trend is clear. CSE and GIBBS are examples of how far a reinvention and integration of education and venture creation can be taken. Some proposals for future steps conclude this investigation.

- How should policies at the national and university levels help the development of such approaches?
- How can this approach be developed to take care of ideas that are even more radical and less low-hanging?
- How can we more readily assess and measure the indirect qualities and values provided by the schools, and the entrepreneurs and ventures they develop?

References

Cohen, B., and Winn, M. I. 2007. Market Imperfections, Opportunity and Sustainable Entrepreneurship. *Journal of Business Venturing* 22 (1): 29–49.

Dean, T. J., and McMullen, J. S. 2007. Toward a Theory of Sustainable Entrepreneurship: Reducing Environmental Degradation through Entrepreneurial Action. *Journal of Business Venturing* 22 (1): 50–76.

Halpern, J., and Plano, L. 2006. $125,000 Ignite Clean Energy Business Presentation Competition Under Way; Invention to Venture Workshop to Train Entrepreneurs February 24. *Business Wire* (February 26): 1.

Innovating Regions in Europe. 2008. Chalmers School of Entrepreneurship, CSE: The Aim of CSE Is to Educate Future Entrepreneurs by Encouraging Students to Pursue Technological Business Ideas. http://www.innovating-regions.org/schemes/scheme.cfm?publication_id=3115&display=byTopic (accessed February 28, 2008).

Marsden, T., and Smith, E. 2005. Ecological Entrepreneurship: Sustainable Development in Local Communities through Quality Food Production and Local Branding. *Geoforum* 36 (4): 440–451.

Pretorius, M., Nieman, G., and van Vuuren, J. 2005. Critical Evaluation of Two Models for Entrepreneurial Education: An Improved Model through Integration. *International Journal of Educational Management* 19 (4/5): 413–28.

CHAPTER 4

Limits of Shareholder Value to Achieving Global Sustainability

Frank Figge and Tobias Hahn

Overview

It is often argued that sustainability and shareholder value are not in conflict with each other but that the goal of sustainable development is in line with the goal of maximizing shareholder value. Using our sustainable value approach we show that corporate sustainability performance can be measured in monetary terms, that sustainability and shareholder value can be in line, but that there is no unambiguous link between the two. Using our database of sustainable value calculations, we give examples of companies that create shareholder value and contribute positively to sustainable development as well as of companies that create shareholder value at the expense of sustainable development. This chapter both provides a methodological approach and makes an empirical contribution to the theory of sustainable development.

1. Introduction

In the conventional view, companies use resources to create a return for their shareholders. In standard management theory the resource economic capital is the focus of attention. As there is a limited amount of economic capital, companies must use this resource efficiently. The simple rule is that the value of a company is the higher the more efficiently capital is being used. This rule is also at the heart of the respective assessment techniques. To calculate shareholder value, the value of the company from the perspective of its shareholders, the efficiency of the capital use of the company is compared to the efficiency of capital use of a benchmark. When the former exceeds the latter, shareholder value is created.

It is increasingly being recognized that company valuation should go beyond the use of economic capital and must also cover environmental and social aspects in addition to economic performance. There are a multitude of different approaches to measure the sustainability performance of companies (Figge 2000; Krajnc and Glavic 2005; Labuschagne, Brent, and Van Erck 2005; Szekely and Knirsch 2005). We distinguish between two kinds of approaches in this context (Figge and Hahn 2008). There are, on the one hand, approaches that aim to cover the environmental, social, and economic performance of companies. All other things being equal, these approaches aim to give a better rating to companies that draw less on environmental, social, and economic resources as compared with those that use more of these resources. On the other hand are approaches that focus on the use of economic capital. These approaches implicitly acknowledge that environmental and social aspects can have an impact on the efficiency of economic capital use (Blumberg, Korsvold, and Blum 1997; Figge 2005; Hart and Milstein 2003). They therefore aim to assess whether companies are more successful in economic terms because of the way they deal with environmental and social aspects.

Both approaches would yield the same results if the following simple truth existed: It always pays to act environmentally and socially responsible. However, it has been argued that such a view on the relationship between environmental and social performance on the one hand and economic performance on the other is simplistic: "Only the most naïve (or blindly hopeful) among us will assume that poor (good) social behavior will always have negative (positive) financial implications" (Rowley and Berman 2000, p. 406). Rather, there is the need to distinguish between different cases and conditions associated with the relationship between environmental, social, and economic performance: "If divorced from their economic context, discussions of business and the environment are too often derailed into sterile arguments about whether it 'pays to be green,' as though the answer had to be categorical"(Reinhardt 1999, p. 10).

By building on this line of argument, we present a measurement approach that allows for the analyses of the relationship between environmental, social, and economic performance in a more sophisticated way. In this context, we make the general assumption that companies can make a positive contribution to sustainable development by using resources efficiently. We emphasize that this covers, but is not limited to, economic resources. Put differently, even a company that uses economic capital inefficiently can make a valid contribution to sustainable development depending on the way it uses the remaining resources.

To examine this relationship between environmental and social performance on the one hand and economic performance on the other, we must be able to measure environmental and social performance and economic performance respectively in a comparable way. It is only then that we will be able to discern whether there is a conflict between the three performance domains or whether, for example, an economic outperformance warrants an environmental and social underperformance. Unfortunately, environmental and social performance on the one hand and economic performance on the other are recorded in different units. A direct comparison is therefore difficult. There are, however, some similarities between the challenges we face when we measure economic, environmental, and social performance of companies. As we will show, it is these similarities that allow us to measure both economic performance and environmental and social performance in the same unit. For this purpose we propose to use the sustainable value approach (Figge 2001; Figge and Hahn 2005).

The remainder of this chapter is structured as follows. In the following section we will look at the similarities and dissimilarities of measuring economic, environmental, and social performance. Section 3 looks at how economic performance is being measured today. In section 4 we show that environmental and social performance can be measured in an analogous manner. Section 5 synthesizes the preceding two chapters by opposing the economic appraisals of section 3 to the environmental appraisals of section 4. The conclusions and implications are discussed in section 6.

2. Similarities and Dissimilarities of the Measurement of Economic, Environmental, and Social Performance

Companies are using economic, environmental, and social resources to create value. To create value the use of a resource must at least cover its cost. What sounds like a simple equation at first has puzzled researchers and practitioners in the field of both economic performance and environmental and social performance because economic, environmental, and social capital are resources without a price tag. Pricing the use of these different forms of capital was therefore considered difficult.

It is sometimes forgotten that there has been an intensive debate about how to determine the cost of economic capital (see, e.g., Warne 1919). There have been diverging opinions about what constitutes a fair price or interest for the use of economic capital. The use of opportunity cost as cost of capital is today considered to be the state-of-the-art solution to this pricing puzzle (Brealey, Myers, and Allen 2008). As will be explained below, the popular shareholder value approach is based on an opportunity cost logic. The return created through an alternative use of a resource constitutes its opportunity cost. A company that aims to increase its shareholder value must therefore increase the efficiency of its use of economic capital relative to its benchmark.

Interestingly, the environmental and social dimension of sustainable development can also be interpreted using a capital approach (Harte 1995; Prugh et al. 1999; Stern 1997). The notion of sustainable development is based on the conviction that society requires not only economic capital but also environmental and social capital to satisfy the needs of this and future generations. From a sustainability point of view the challenge is that all forms of capital are scarce and some forms of capital are being overused. Similar to the use of economic capital, a more efficient use of environmental and social capital is therefore being called for. Different names and buzz-words are used in this context. Factor X, Factor 4, and Factor 10 are examples (Schmidt-Bleek and Weaver 1998).

There is widespread agreement that increasing the efficiency of resource use is a necessary condition for achieving sustainability. However, it is not sufficient. This is mainly due to two reasons. Increasing the efficiency of resource use can, on the one hand, lead to a higher overall use of capital. This is known as the rebound effect (Berkhout, Muskens, and Velthuijsen 2000; Greening, Greene, and Difiglio 2000). The rebound effect can, for example, occur when a higher efficiency results in lower prices, which will in turn lead to a higher demand that overcompensates the resources saved. On the other hand, it is important to look at the relationship between the different forms of capital. Companies as well as society at large require a mix of different forms of capital. There are therefore limits to the substitution of one form of capital (e.g., economic capital) for another (e.g., environmental capital). In this context, it is common to distinguish between weak forms of sustainability, which allow for substitution between different forms of capital, and strong forms of sustainability, which limit the substitution of different forms of capital (Neumayer 1999).

Depending on the position that we assume—that is, whether we want to allow for the substitution of different forms of economic capital or not—we must make sure not only that resources are being used efficiently but also that the stocks of different forms of capital are being preserved. The rebound effect is usually not discussed in the context of the use of economic capital, because economic capital is reproducible and an overuse of economic capital is, in contrast to environmental and social capital, not a matter of concern.

There is a great range of approaches that aim to price the use of environmental and social capital (see, e.g., Costanza et al. 1997; Richmond, Kaufmann, and Myneni 2007). While they differ in detail they all use a similar way of thinking to measure environmental and social performance. Environmental and social performance is measured on the basis of the burden that it poses on the environment and on society (Figge and Hahn 2004b). Methods like the ecological footprint are a case in point (Chambers and Lewis 2001; Wackernagel and Rees 1996). This method translates the use of environmental resources into units of area used. A resource that uses up twice as much area than another resource will also be twice as burdensome.

Interestingly, attempts to measure the cost of capital on the basis of a burden-based logic were abandoned in corporate finance a long time ago. The measurement of economic performance follows a value-based logic. This step has only recently been taken with regard to environmental and social performance (Figge 2001; Figge and Hahn 2004a, 2005; Hahn, Figge, and Barkemeyer 2007).

3. The Shareholder Value Approach

Many practitioners as well as management researchers consider shareholder value the standard method to measure corporate financial performance. There are a number of different approaches to measure shareholder value (Copeland, Koller, and Murrin 2000; Rappaport 1986). Shareholder value is defined as the net present value of the funds that shareholders may expect to receive in the future after the cost of capital has been covered. What all these approaches have in common is that they use the concept of opportunity costs (Bastiat 1870) to determine the cost of capital. Put differently, shareholder value is created when the return created by using economic capital exceeds the return that would have been created by an alternative use of economic capital. In the context of this chapter we make two simplifications. First, we compare the use of resources across a single period. Second, we do not distinguish between different levels of risk—that is, we assume that all companies are exposed to the same level of risk. Both simplifications are only made to keep the underlying argument clear. They do not limit the explanatory power of the argument.

If we assume that the return a company creates is measured by its operating profit and the capital employed by a company is measured by its total assets, then the return on capital can be defined as the operating profit per total assets. A company will create shareholder value when it creates more operating profit with its economic capital than the benchmark would have created. Put differently, the company's return on capital must be above other companies' return on capital. Shareholder value is therefore created when the following holds true:

$$\frac{\text{Operating Profit}_C}{\text{Total Assets}_C} > \frac{\text{Operating Profit}_B}{\text{Total Assets}_B}$$

where Operating Profit$_C$ is the operating profit of the company; Total Assets$_C$, assets used by the company; Operating Profit$_B$, the operating profit of other companies; and Total Assets$_B$, assets used by other companies.

To create shareholder value it is neither sufficient nor necessary that a company creates a positive return on capital. Shareholder value is created when the company's return on capital exceeds that of a benchmark. Shareholder value is therefore a relative concept. The benchmark performance can of course be negative, in which case even a negative return on capital can result in the creation of shareholder value as long as it exceeds the return on capital of the benchmark.

To determine the contribution to shareholder value, one must first calculate the value spread by subtracting the benchmark's return on capital from that of the company. A positive value spread shows that there is shareholder value creation. To calculate the absolute contribution to shareholder value, in a second step the value spread is multiplied with the capital employed by the company. This shows by how much the company has contributed to shareholder value. A company's contribution to shareholder value is thus equivalent to the additional return that is created by the use of economic capital relative to a situation in which the economic capital had been used by other companies.

The contribution of economic capital to shareholder value for BMW is calculated in figure 4.1. In this case shareholder value is calculated relative to other car manufacturers.

As we can see in this example, BMW had a return on capital of about 5 percent in 2005. The other car manufacturers only had a return of about 2.5 percent. BMW therefore created about twice as much return with its economic capital than its peers. BMW's overall contribution to shareholder value in 2005 can therefore be estimated at almost €2 billion.

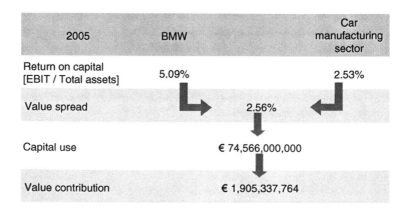

Figure 4.1 Calculating contribution to shareholder value (based on company reports)

4. The Sustainable Value Approach

When we measure the sustainability performance of companies we face a familiar problem. Putting a price tag on environmental and social resources used is difficult, if not impossible, in practice. There has been substantial research on costing the use of environmental and social resources (Costanza et al. 1997; Herendeen 1998). All approaches used to date follow a similar logic: the cost of a resource is equated to the cost, pain, or burden its use inflicts on society. While methodologically sound, these approaches hit considerable problems when applied in practice. It is impossible in practice to arrive at a consensus regarding the burden-related costs of all environmental, social, and economic resources. As mentioned above, economics has faced a similar problem with regard to the cost of economic capital. Determining the right or fair interest for the use of economic capital was difficult and subject to much debate (Warne 1919). The solution now used in corporate finance is to price economic capital using opportunity cost thinking.

The sustainable value approach (Figge 2001; Figge and Hahn 2004a, 2005; Van Passel, Nevens, Mathijs, and Van Huylenbroeck 2007) adopts this value-oriented approach to the pricing of resources and widens it to also cover environmental and social resources. To create sustainable value the use of a resource must earn its opportunity cost. For this reason the efficiency of a company's resource use is compared with the efficiency of a benchmark's resource use. Sustainable value is created when a company uses a resource more efficiently than a benchmark does. The benchmark can, for example, be an entire economy or other companies of a sector.

Figure 4.2 conducts an analogous assessment to figure 4.1. Rather than economic capital, we now assess whether carbon dioxide (CO_2) was used in

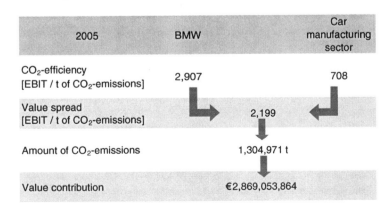

Figure 4.2 CO_2-value contribution of BMW in 2005 (based on company reports)

a value-creating way. For this purpose, the efficiency of BMW's CO_2 use is compared with the efficiency of the car-manufacturing sector's CO_2 use.

As we can see in figure 4.2, in 2005 BMW created €2,907 EBIT (earnings before interest and taxes) per ton of CO_2, while the car-manufacturing sector created only about €708. There is an additional €2,199 for every ton of CO_2 that is used by BMW rather than by the benchmark. Since BMW emits a total of 1,304,971 tons of CO_2 an excess return of €2,869,053,864 results. This value contribution can be interpreted analogously to a company's contribution to shareholder value.

Automobile manufacturers use, of course, not only CO_2 but many different resources. In the financial markets it is implicitly assumed that the entire return is created by a single resource—economic capital. Consequently, the entire return is related to this one resource. When calculating eco-efficiency the same simplification is made. Obviously, companies require more than a single resource to create a return. When we relate each single resource to the entire return, there is therefore double counting. To take into account this double-counting effect we must determine the relative weight of the different resources.

In this context, burden-oriented approaches determine the relative cost, burden, or harmfulness of the use of one resource compared with another resource. A popular approach in this context is, for example, to determine the relative harmfulness of one greenhouse gas (e.g., CO_2) relative to another greenhouse gas (e.g., CH_4). As mentioned earlier these approaches run into considerable problems in their practical application. This is, for example, the case when two emissions contributing to different environmental problems are to be compared. It is useful to remember why we are using economic, environmental, and social resources in the first place. "The value of an object is not derived from the sacrifice made to obtain it. On the contrary we make the sacrifice because the object has this value. The value is first, the sacrifice second" (Whitaker 1904, p. 74). Sustainable value is a value-oriented approach. Consequently, resources are weighted according to their value contribution. We assume that the relative contribution of one resource relative to another resource is defined by its eco-efficiency relative to the other resource's eco-efficiency. A resource that is used twice as efficiently compared with another resource on the benchmark level will therefore also have twice the weight.

So far, and for the sake of explaining the fundamental notion of the sustainable value approach, we have focused on the use of just one resource. Following the notion of sustainable development, sustainable value usually analyzes the use of a bundle of economic, environmental, and social resources. As an indicator of corporate sustainability performance, sustainable value thus reflects the excess return that is created by an entire set of economic, environmental, and social resources. In other words it reflects the extra return that is created because overall a set of resources is used more efficiently by the company than by the benchmark. In the underlying model the overall resource use is kept constant on the benchmark level because the approach does not venture to answer the question whether a resource should be used but whether the resource should be used by the company or by the benchmark. By adjusting the overall amount of resources available on the benchmark level when defining the benchmark eco-efficiency, limits on the overall use of environmental and social resources can be set. The approach can therefore be used in conformity with different levels of weak and strong sustainability.

5. Sustainable Shareholder Value?

There is a wealth of literature that aims to link environmental and social performance to the generation of shareholder value (Figge 2005; Hart and Milstein 2003; Repetto and Austin 2001; Vafeas, Nikolaou, and Cheryl 2001). The underlying message of this literature is that companies can create shareholder value by improving their environmental and social performance. There are no doubt many examples that show that these so-called win-win cases or business cases for sustainability do exist. Companies can increase shareholder value by improving their environmental and social performance. What is of interest is the question whether aiming for higher shareholder value will also (tend to) increase environmental and social performance. Are companies that create more shareholder value per se more sustainable than companies that create less shareholder value? If that was the case, would it be sufficient to make sure that companies strive to maximize their shareholder value to, at the same time, arrive at a higher environmental and social performance? Moreover, following the logic of the business case for sustainability implies that corporate environmental and social performance is only relevant to the degree that it contributes to a higher return on economic capital. This represents a hierarchical subordination of environmental and social aspects under economic performance (Figge and Hahn 2008) and is thus not in line with the notion of sustainable development (Dyllick and Hockerts 2002). There are thus two questions that are relevant in this context that address the relationship between economic, environmental, and social performance from two opposing angles: Does a high economic performance (i.e., the strive toward shareholder value) also result in a higher corporate environmental and social performance? And does a good environmental and social performance contribute to a superior economic performance? Both questions have been addressed by numerous empirical studies that correlate economic, environmental, and social performance (see, e.g., Clarkson, Li, Richardson, and Vasvari 2007; McWilliams and Siegel 2000; Montabon, Sroufe, and Narasimhan 2007). However, we argue that the underlying notion of the so-called business case is not sufficient to address this complex relationship and that there is a need for a more sophisticated analysis of the sustainability performance of companies.

We argue that it is fruitful to follow a value-oriented paradigm for such an analysis. As shown above, using the sustainable value approach to assess corporate sustainability performance helps to tackle the problem of pricing environmental and social resources analogously to the way economic capital is being valued on financial markets today. In the following we develop how the relationship between corporate economic, environmental, and social performance can be fruitfully analyzed applying opportunity cost thinking not only to economic capital but also to environmental and social resources.

For the time being and for developing the underlying logic of this argument, we limit ourselves to economic capital to represent economic performance and CO_2 emissions as a proxy for environmental performance. On the basis of these two resources we can come up with a matrix (figure 4.3) that shows the different possible combinations of economic and environmental performance. Using the valuation logic developed above we can determine for every resource if it contributes to value creation. This builds on the shared characteristics of the shareholder value and the sustainable value approach. According to both approaches value is created whenever the use of the resource yields more return than on the benchmark level. Figure 4.3 addresses a simple case where there is one economic and one environmental resource. It is determined whether each resource is being used in a value-creating way—that is, if there is a positive contribution to economic performance (positive contribution to shareholder value) and a positive contribution to environmental performance (positive contribution to environmental value[1]).

If a high (or low) environmental performance was always linked to the creation (or destruction) of shareholder value, then all companies should be placed in either the upper-right-hand or the lower-left-hand field of the matrix. As already argued in the introduction, this has been identified as a simplistic view. Although it may be simplistic, it is still at the heart of how resources are allocated today in corporate practice. The environmental and social aspects and the use of economic capital are assessed and managed with regard to their contribution to a higher return on economic capital. It is implicitly assumed that companies that are using their economic capital

		Creation of shareholder value & destruction of environmental value	Creation of environmental value & shareholder value
Economic performance	Positive value contribution		
	Negative value	Destruction of environmental & shareholder value	Destruction of shareholder value & creation of environmental value
		Negative value contribution	Positive value contribution
		Environmental performance	

Figure 4.3 Sustainable Value Matrix

efficiently will thus flourish in the marketplace and will also use environmental resources efficiently. With the help of the sustainable value matrix we can show that this is not necessarily the case. At worst, the selection of companies on the basis of their economic efficiency can even lead to a situation in which the overall efficiency of resource uses degrades. Inversely, this thinking does not capture situations in which a company has a poor economic performance but contributes widely to environmental and social value creation.

To demonstrate this effect we look at the economic and environmental performance of automobile manufacturing. We use the return on economic capital (EBIT per total assets) and the return on CO_2 (EBIT per ton of CO_2) as indicators of economic and environmental performance. EBIT has been adjusted for exceptional items. To calculate the opportunity cost we determine the average performance in the car-manufacturing sector over the five-year period between 2001 and 2005.[2] During this period the automobile sector created on average €2.97 EBIT per €100 of economic capital. To contribute to shareholder value creation relative to their peers in that year, companies therefore needed to attain a return on assets of at least 2.97 percent. We can determine the opportunity cost of CO_2 analogously. In the five-year period between 2001 and 2005 the automobile sector created €784 EBIT per ton of CO_2 on average. To contribute to the creation of environmental value within the car-manufacturing sector a company must have a CO_2 efficiency of more than €784 per ton of CO_2 emissions. We can now determine which companies have used both their economic capital and their CO_2 emissions in a value-creating way. We illustrate the four different cases sketched out above by filling the four fields of the matrix of figure 4.4 with real-world examples.

Economic performance	Positive value contribution	Daihatsu	BMW
	Negative value contribution	GM	Renault
		Negative value contribution	Positive value contribution
		Environmental performance	

Figure 4.4 Sustainable Value Matrix of car manufacturers

BMW is a good example of a company that has used both resources, economic capital and CO_2, in a value-creating way. In the period between 2001 and 2005 BMW had an EBIT return on economic capital of about 5.8 percent and generated an EBIT of about €3,231 per ton of CO_2 on average. BMW is therefore placed in the upper-right-hand corner of the matrix (fig. 4.4).

The French carmaker Renault is a good example of a company with a positive environmental and a negative economic performance. Renault achieved a return on economic capital of 2.4 percent and a return on CO_2 of about €1,950 per ton of CO_2 during the same period. Renault is therefore placed in the lower-right-hand corner of the matrix.

Daihatsu is an example of a company that is using its economic resources in a value-creating way (3.7 percent) while not being able to reach the market return on CO_2 (€626 per ton of CO_2). Daihatsu is therefore placed in the upper-left-hand corner of the matrix.

There are several companies that fail to meet the economic and environmental threshold efficiencies. GM is a good example with an economic performance of only 0.4 percent and €111 EBIT per ton of CO_2. GM is therefore placed in the lower-left-hand corner of the matrix.

As this matrix shows, there are real-world examples for all four cases. There are two clear-cut cases. BMW contributes to a more efficient use of economic and environmental resources relative to the sector and GM fails to keep up with its peers. The interpretation of the performance of Renault and Daihatsu is more complicated because they outperform the sector in one dimension and underperform in the other dimension. To be able to compare the effect of the overperformance with the effect of the underperformance, one can compare the overall value contribution of the two dimensions.

The value contribution reflects how much more return was created because the resource was used by the company rather than by its peers.

In the case of Daihatsu we know that the total assets amount to about €6 billion. Daihatsu's return on assets is about 0.72 percent higher than the return on assets of the sector. The use of economic capital therefore resulted in a value contribution of about €44 million (figure 4.5). At the same time, the company used CO_2 less efficiently than the market. The market created €784 EBIT per ton of CO_2 compared with €626 of the company. There was therefore an underperformance of €158 per ton of CO_2. As Daihatsu used 363,600 tons on average, an estimated €57 million was lost.

Both the economic and the environmental performance are now expressed in monetary terms. A comparison of the value contribution shows that the economic outperformance is inferior to the environmental underperformance. This is due to the fact that the amount of excess return we have obtained from giving economic capital to Daihatsu rather than to its peers is inferior to the return that we have lost for giving CO_2 to Daihatsu rather than to the sector on average.

An analogous analysis of the French carmaker Renault shows that there is an environmental outperformance of about €835 million, which is strong enough to compensate for the economic underperformance of about €332 million (figure 4.6).

If we assume that both economic capital and CO_2 constitute a scarce resource and that both resources are required for companies to operate, then it becomes clear that the overall resource efficiency (i.e., including both the use of economic capital and CO_2) of Renault lies above the market level, while the overall efficiency of Daihatsu does not keep up with its sector peers.

As the examples of Renault's and Daihatsu's performance illustrate, assessing corporate sustainability performance following a value-oriented paradigm offers insights that go beyond categorical assumptions on the relationship between economic, environmental, and social performance. Rather, this

	Company efficiency	Sector efficiency	Resource input	Value contribution
Economic performance	(3.69% -	2.97%) *	€ 6,083,669,984 =	€ 43,802,424
Environmental performance	(626 €/t of CO_2 -	784 €/t of CO_2) *	363,600 t of CO_2 =	-€ 57,448,800

Figure 4.5 Performance of Daihatsu

	Company efficiency	Sector efficiency	Resource input	Value contribution
Economic performance	(2.40% -	2.97%) *	€ 58,200,200,000 =	-€ 331,741,140
Environmental performance	(1,950 €/t of CO_2 -	784 €/t of CO_2) *	716,124 t of CO_2 =	€ 835,000,584

Figure 4.6 Performance of Renault

paradigm offers a more differentiated analysis that allows us to evaluate conflicting situations where there is a trade-off between corporate economic, environmental, and social performance. This has some interesting conceptual and practical implications, which are addressed in the next section.

Implications and Conclusions

Companies compete for the use of economic capital in the marketplace. Companies that are able to at least match the return on capital of the market will be able to attract economic capital. Companies that fail to offer an equivalent return will be under pressure to strive for a more efficient use of capital and will be pushed off the market in the long run. It is being claimed that this is a virtuous process that results in a better or even optimal allocation of economic capital (Bughin and Copeland 1997). More efficient companies will dominate less efficient companies. Companies will therefore strive for efficiency. Companies that are lagging behind in terms of efficiency will embrace the tools and techniques proposed by the leaders. Put differently, efficiency leaders will become the role models for efficiency laggards. The champions of the business case for sustainability see this as an opportunity for sustainability.

If one follows the notion of sustainable development, it becomes clear that companies need not only economic capital but also environmental and social resources to create a return. There is a complementary relationship between these resources. Economic resources cannot create a return without environmental resources, and vice versa. On the financial markets a company's license to operate is, however, primarily determined by its ability to generate above-average returns on economic capital. A company that fails to produce a return on economic capital that is at least in line with the market is likely to disappear. Companies must at least cover their opportunity costs of economic capital in the long run.

The amount of environmental and social resources being used to produce this return will only be considered indirectly. For some environmental and social resources there are prices. CO_2 is a good example in this context. CO_2-emissions are mostly linked to the use of energy. The cost of energy can therefore be considered a price for the use of CO_2 even when there is no market for CO_2-emissions. Using more environmental resources will then lead to higher costs and consequently a lower return on economic capital. Emissions will therefore be reduced as long as the cost of reducing the emissions is lower than the cost of the emissions. Put differently, emissions will be reduced as long as this contributes to the creation of shareholder value. This will not necessarily lead to higher eco-efficiency. Investors who decide where to invest might choose a CO_2-intensive company. This can be illustrated using the example of the four car companies above. Investors are likely to favor Daihatsu over Renault as Daihatsu offers a higher return on economic capital. However, as Daihatsu has a lower CO_2 efficiency than Renault this will result in a decrease of eco-efficiency in this sector.

The argument developed in this chapter offers more detailed insights into the relationship between corporate economic, environmental, and social performance. Most importantly this refers to the two main questions introduced above: Does the striving for shareholder value bring about improvements in environmental and social performance? And does higher environmental and social performance deliver better financial results? On a more general level our argument questions the dominating paradigm of the business case for corporate sustainability from a conceptual point of view in that it clearly shows the limits of the explanatory power that can be expected from such kind of analyses.

The argument developed in this chapter clearly shows the limits of the popular notion of the business case for sustainability. It becomes clear that a business case for sustainability will only exist where the cost of reducing the use of a resource is less than the private cost of emitting this resource. It is in these cases that we may expect a virtuous win-win opportunity. Thus, there is only a limited universe of cases in which the business case applies. Moreover, following the win-win logic one turns a blind eye on cases in which there may be an economic underperformance that is, however, outweighed by an environmental outperformance, as in the case of Renault. Basing both the assessment and the management of corporate sustainability performance on the win-win logic will thus result in a purely economic optimization of corporate environmental and social performance. Environmental and social aspects will only be considered to the degree to which they contribute to a higher level of economic efficiency. This clearly falls short of the requirements of the notion of sustainable development.

What has also become clear is that general and categorical assumptions and statements on the relationship between economic, environmental, and social performance are not valid. This holds for both perspectives, the assertion that the striving toward shareholder value will also yield good environmental and social performance and the belief that good environmental and social performance will result in superior economic performance. Rather, it has become clear that economic, environmental, and social resources are complementary for the generation of profits. A valid approach to the measurement and assessment of corporate sustainability performance needs to capture this fundamental requirement. Consequently, from a sustainable development perspective it follows that corporate sustainability assessments should focus on the return on a bundle of different resources rather than on single aspects or creating hierarchical biases. Most importantly, corporate sustainability analysis needs to be able to address and solve trade-off situations. This becomes particularly clear in the cases of Renault and Daihatsu. Economic outperformance is not a guarantee for an overall efficient use of the bundle of economic, environmental, and social resources, as has become clear by the case of Daihatsu. On the contrary, a strong environmental outperformance may even be sufficient to outweigh an economic underperformance, as in the case of Renault. Both the paradigm of the business case for sustainability and the notion of win-win fall short of solving such trade-off situations.

As this research shows, there is a need to go beyond correlation studies as these are not suitable to capture the complementarity of economic, environmental, and social aspects in sustainability performance. In this chapter we have demonstrated how this problem can be addressed by adopting a value-oriented approach toward the assessment of corporate sustainability performance. Applying opportunity cost thinking to the analysis of economic, environmental, *and* social resources provides a number of advantages in this context. First, it allows us to measure economic, environmental, and social performance in a similar way. Second, it helps us to avoid the problem of determining the damage costs of using environmental and social resources. Third, it allows us to address and evaluate trade-offs and conflicting situations between different aspects and domains of corporate sustainability performance.

If companies are to contribute to a more efficient resource use, they must dispose of a measurement tool that enables them to reliably assess their performance. Following the argument in this chapter it becomes clear that this measurement tool is neither sufficient to strive for a higher return on capital nor reliable to aspire for an automatic link between a good environmental performance and high economic returns. Most interestingly, this does not require a shift away from widely accepted economic notions like opportunity cost thinking. Rather, it was shown that the further development of such notions bears the potential for innovative solutions to the corporate sustainability puzzle.

Acknowledgments

We are indebted to our colleagues Ralf Barkemeyer and Andrea Liesen, who continue to amaze us with their data mining and analytical skills. Financial support from the German Federal Ministry of Education and Research under grant number 07 IFS 11 is gratefully acknowledged.

Notes

1. By speaking of environmental value we refer to the special case in which sustainable value is calculated for just one environmental resource. Alternatively, we could also speak of CO_2-value.

2. The performance of the car manufacturing sector is determined by the weighted average of the performance of the following sixteen car manufacturers: BMW Group, Daihatsu, DaimlerChrysler, Fiat Auto, Ford, GM, Honda, Hyundai, Isuzu, Mitsubishi, Nissan, PSA, Renault, Suzuki, Toyota, and Volkswagen Group.

References

Bastiat, F. 1870. Ce qu'on voit et ce qu'on ne voit pas. In *Oeuvres complètes de Frédérick Bastiat, mises en ordre, revues et annotées d'après les manuscrits de l'auteur*, ed. F. Bastiat, 3rd ed., vol. 5, pp. 336–392. Paris: Guillaumin.

Berkhout, P. H. G., Muskens, J. C., and Velthuijsen, J. W. 2000. Defining the rebound effect. *Energy Policy* 28 (6–7): 425–432.

Blumberg, J., Korsvold, Å., and Blum, G. 1997. *Environmental performance and shareholder value*. Geneva: WBCSD.

Brealey, R. A., Myers, S. C., and Allen, F. 2008. *Principles of corporate finance*. 9th ed. Boston: McGraw-Hill.

Bughin, J., and Copeland, T. E. 1997. The virtuous cycle of shareholder value creation. *The McKinsey Quarterly* (2): 156–167.

Chambers, N., and Lewis, K. 2001. *Ecological footprint analysis: Towards a sustainability indicator for business*. London: The Association of Chartered Certified Accountants.

Clarkson, P. M., Li, Y., Richardson, G. D., and Vasvari, F. P. 2007. Revisiting the relation between environmental performance and environmental disclosure: An empirical analysis. *Accounting, Organizations and Society*. (In press, corrected proof.)

Copeland, T. E., Koller, T., and Murrin, J. 2000. *Valuation. Measuring and managing the value of companies*. 3rd ed. New York: John Wiley & Sons.

Costanza, R., d'Arge, R., de Groot, R., Farber, S., Grasso, M., Hannon, B., Limburg, K., et al. 1997. The value of the world's ecosystem services and natural capital. *Nature* 387 (May): 253–260.

Dyllick, T., and Hockerts, K. 2002. Beyond the business case for corporate sustainability. *Business Strategy and the Environment* 11 (2): 130–141.

Figge, F. 2000. *Öko-Rating. Ökologieorientierte Bewertung von Unternehmen*. Berlin: Springer.

Figge, F. 2001. Environmental value added: Ein neues Maß zur Messung der Öko-Effizienz. *Zeitschrift für Angewandte Umweltforschung* 14 (1–4): 184–197.

Figge, F. 2005. Value-based environmental management. From environmental shareholder value to environmental option value. *Corporate Social Responsibility and Environmental Management* 12 (1): 19–30.

80 • Figge and Hahn

Figge, F., and Hahn, T. 2004a. Sustainable value added: Measuring corporate contributions to sustainability beyond eco-efficiency. *Ecological Economics* 48 (2): 173–187.

Figge, F., and Hahn, T. 2004b. Value-oriented impact assessment: The economics of a new approach to impact assessment. *Journal of Environmental Planning and Management* 47 (6): 921–941.

Figge, F., and Hahn, T. 2005. The cost of sustainability capital and the creation of sustainable value by companies. *Journal of Industrial Ecology* 9 (4): 47–58.

Figge, F., and Hahn, T. 2008. Sustainable investment analysis with the Sustainable Value Approach: A plea and a methodology to overcome the instrumental bias in socially responsible investment research. *Progress in Industrial Ecology* 5 (3): 255–272.

Greening, L. A., Greene, D. L., and Difiglio, C. 2000. Energy efficiency and consumption: The rebound effect—a survey. *Energy Policy* 28 (6–7): 389–401.

Hahn, T., Figge, F., and Barkemeyer, R. 2007. Sustainable Value creation among companies in the manufacturing sector. *International Journal of Environmental Technology and Management* 7 (5/6): 496–512.

Hart, S. L., and Milstein, M. B. 2003. Creating sustainable value. *Academy of Management Executive* 17 (2): 56–69.

Harte, M. J. 1995. Ecology, sustainability, and environment as capital. *Ecological Economics* 1995 (15): 157–164.

Herendeen, R. A. 1998. Monetary-costing environmental services: Nothing is lost, something is gained. *Ecological Economics* 25 (1): 29–30.

Krajnc, D., and Glavic, P. 2005. How to compare companies on relevant dimensions of sustainability. *Ecological Economics* 55 (4): 551–563.

Labuschagne, C., Brent, A. C., and Van Erck, R. P. G. 2005. Assessing the sustainability performances of industries. *Journal of Cleaner Production* 13 (4): 373–385.

McWilliams, A., and Siegel, D. 2000. Corporate social responsibility and financial performance: Correlation or misspecification? *Strategic Management Journal* 21 (5): 603–609.

Montabon, F., Sroufe, R., and Narasimhan, R. 2007. An examination of corporate reporting, environmental management practices and firm performance. *Journal of Operations Management* 25 (5): 998–1014.

Neumayer, E. 1999. *Weak versus strong sustainability: Exploring the limits of two opposing paradigms.* Cheltenham: Edward Elgar.

Prugh, T., Costanza, R., Cumberland, J. H., Daly, H. E., Goodland, R., and Norgaard, R. B. 1999. *Natural capital and human economic survival.* 2nd ed. Boca Raton: Lewis.

Rappaport, A. 1986. *Creating shareholder value. The new standard for business performance.* New York: The Free Press.

Reinhardt, F. 1999. Market failure and the environmental policies of firms. Economic rationales for "beyond compliance" behavior. *Journal of Industrial Ecology* 3 (1): 9–21.

Repetto, R., and Austin, D. 2001. Quantifying the impact of corporate environmental performance on shareholder value. *Environmental Quality Management* 10 (4): 33–44.

Richmond, A., Kaufmann, R. K., and Myneni, R. B. 2007. Valuing ecosystem services: A shadow price for net primary production. *Ecological Economics* 64 (2): 454–462.

Rowley, T., and Berman, S. 2000. A brand new brand of corporate social performance. *Business & Society* 39 (4): 397–418.

Schmidt-Bleek, F., and Weaver, P. 1998. *Factor 10. Manifesto for a sustainable planet.* Sheffield: Greenleaf.

Stern, D. I. 1997. The capital theory approach to sustainability: A critical appraisal. *Journal of Economic Issues,* 31 (1): 145–173.

Szekely, F., and Knirsch, M. 2005. Responsible leadership and corporate social responsibility: Metrics for sustainable performance. *European Management Journal* 23 (6): 628–647.

Vafeas, N., Nikolaou, V., and Cheryl, R. L. 2001. The association between corporate environmental and financial performance. *Advances in Public Interest Accounting* 8: 195–214.

Van Passel, S., Nevens, F., Mathijs, E., and Van Huylenbroeck, G. 2007. Measuring farm sustainability and explaining differences in sustainable efficiency. *Ecological Economics* 62 (1): 149–161.

Wackernagel, M., and Rees, W. E. 1996. *Our ecological footprint. Reducing human impact on the Earth.* Gabriola Island, BC: New Society.

Warne, F. J. 1919. Corporation finance and the wage worker. *Annals of the American Academy of Political and Social Science* 85 (1): 271–278.

Whitaker, A. C. 1904. *History and criticism of the labor theory of value in English political economy.* Columbia University, New York.

CHAPTER 5

Green Chemistry and EVA: A Framework for Incorporating Environmental Action into Financial Analysis

Geoff Archer, Andrea Larson, Mark White, and Jeffrey G. York

> *What we have here is a failure to communicate.*
> Captain, Road Prison 36,
> *Cool Hand Luke* (1967)

Introduction

The past few decades have seen increased interest and concern around businesses' relationship with their natural environments and, in particular, ways in which businesses can either help or hinder the transformation to a sustainable society (Hart 2005; Hawken, Lovins, and Lovins 2000; Hoffman 2000; Reinhardt 2000). Many frameworks and decision-making tools, most of which originated in engineering or product design communities, have been developed to address these issues. While these tools have recently come to the attention of corporate strategists and financial managers, widespread adoption does not seem imminent and is pending exposition of a compelling "business case."

Like the prison guard in *Cool Hand Luke*, we believe that a key element delaying greater integration of sustainability and strategic and financial concerns is a "failure to communicate"—a disconnect between plant managers, product designers, and financial decision makers. Of particular concern is a lack of communication between those knowledgeable about the health and environmental impacts of business and the people making decisions about corporate investments and strategic direction. To remedy this situation, this chapter describes the potential economic benefits of pursuing a specific manufacturing/design paradigm known as "green chemistry." The goal of green chemistry is to reduce the amount of materials employed, and particularly the levels of hazardous substances used or emitted in the production of manufactured goods.

Our work uses economic value added (EVA), a well-respected financial framework, to illustrate the benefits of pursuing a green chemistry strategy. The result is a two-dimensional matrix that we hope will provide grounds for conversation between the manufacturing/design and finance/strategy communities. Fruitful opportunities for sustainability-oriented collaboration are shown to exist regardless of one's starting perspective—that is, engineering/chemistry or finance/strategy. Points of overlap in the matrix offer an expedient way to assess how decisions can simultaneously and positively influence financial and environmental returns. This kind of analysis can be extended to other frameworks to provide important tools to advance both agendas.

The organization of this chapter is as follows: First, we briefly review the concept of sustainability and describe various frameworks and tools that have been used to advance it. Next, we focus on green chemistry, a particularly promising field of inquiry that, we believe, holds significant promise in addressing fundamental manufacturing challenges. We then move to a discussion of EVA and its use in financial decision making. The heart of our work is an analysis highlighting the financial and strategic implications of each of the twelve principles of green chemistry. The chapter concludes with a discussion and summary of our findings.

Sustainability Frameworks

"Sustainability" is a rather slippery term generally intended to convey a concern for the future persistence of humankind. This modern usage first surfaced as a concept during the 1987 World Conference on the Environment held in Stockholm. Gro Harlem Brundtland, the then prime minister of Norway, was responsible for writing the conference's final report, *Our Common Future*, which defined a sustainable society as "a society that meets the needs of the present without compromising the ability of future generations to meet their own needs" (World Commission on Environment and Development 1987).

We interpret "business sustainability" as a commitment by firms to give simultaneous consideration to social, environmental, health, and economic performance criteria with the goal of achieving a society of permanence.

A variety of conceptual frameworks have arisen to aid practitioners in simplifying, quantifying, and communicating advances toward sustainability. A few of the more prominent ones are summarized in the following sections.

Industrial Ecology

Industrial ecology offers a systems approach to understanding the interaction between industry and the natural world. It looks beyond the linear "cradle-to-grave" viewpoint of conventional business design and imagines industry as a series of material and energy flows in which the wastes of one process ideally serve as the feedstock of another, mimicking cycles of nature where there is no waste. The observation that "materials and energy are the interdependent feedstocks of economic systems, and thermodynamics is their moderator" (Cloud 1997) spread with the coining of the term "industrial ecology" by Robert A. Frosh and Nicholas Gallopoulos (1989).

Industrial ecology concepts have been used and popularized more recently by Hawken (1993) and Braungart and McDonough (2002). Several common and key elements are present, such as the systems view, the importance of flows and relationships rather than isolated components, and the cradle-to-cradle (C2C) design philosophy. Industrial ecology is grounded explicitly, however, on the scientific principles of materials and energy flows.

Eco-Efficiency

Another concept that has evolved into a sustainability framework is *eco-efficiency*, or essentially "doing more with less," that is, providing more products or services while using fewer resources and generating less waste (Schmidheiny 1992). A commitment to eco-efficiency lies at the heart of many sustainability frameworks, for example, design for the environment, green chemistry, and green engineering, and the concept is quickly grasped by most businesspersons. Despite its obvious merit, eco-efficiency has its critics, most notably William McDonough (Stromberg 2002), on the grounds that it begs the "goodness" question.

McDonough (Stromberg 2002) argues that one should strive for eco-effectiveness—"doing the right thing." Additionally, Larson (Larson 2000; Larson, Teisberg, and Johnson 2000) argues that eco-efficiency is a necessary but not sufficient condition for movement toward a more cyclically oriented and therefore sustainable economy. Eco-efficiency practices are critical to dematerialization efforts and often the stimulus to innovation, or what McDonough and Braungart (2002) call eco-effectiveness.

Life Cycle Analysis

Life cycle analysis (LCA) refers to the compilation, categorization, and evaluation of the inputs, outputs, and potential environmental impacts of a product, process, or system from raw materials extraction to disposal ("cradle to grave"). A complete LCA is made up of three stages:

1. Definition of goals and scope
2. Life cycle inventory of energy and materials used at each stage of the cycle
3. Life cycle assessment of the environmental impact of inputs and outputs at each stage

The implicit goal of life cycle analysis is product optimization. LCAs provide comparative data that assist producers and manufacturers to select processes and develop products with the least environmental impacts. For example, which is more environmentally benign? Recyclable, yet heavy, glass bottles that must be transported to and from distribution centers, or disposable, lightweight plastic bottles manufactured from nonrenewable petroleum? (The answer is, "It depends.") LCA is often an important precursor for decision making in other sustainability frameworks.

Cradle to Cradle

In their book *Cradle to Cradle*, the architect William McDonough and the chemist Michael Braungart (2002) proposed three principles for ecological design that provide a framework for sustainable investment decisions. Perhaps the best known of these is "waste equals food," a clever reference to natural food chains, in which organisms of one trophic level provide nutrition for the next. McDonough and Braungart's other design principles, "use current solar income" and "celebrate diversity," are also inspired by natural systems and share commonalities with the green chemistry and green engineering frameworks described later (McDonough, Braungart, Anastas, and Zimmerman 2003).

In the C2C framework, technical nutrients (metals, plastics, etc.) are recycled into new products, while biological nutrients (organic materials) are returned to the soil for regeneration.

The Natural Step

In 1989, Karl-Henrik Robèrt, a Swedish physician concerned about increasing cancer rates in children, wrote a paper elucidating the scientific principles needed to ensure a healthy, sustainable society. He established a foundation, *The Natural Step*, to promote and support the implementation of this framework (K.-H. Robèrt 2002a; K.-H. Robèrt 2002b; Robèrt et al. 2002). Today *The Natural Step* operates in nine countries and has assisted more than one hundred firms and municipalities in the transition to sustainability.

According to *The Natural Step*, in a sustainable society, nature is not subject to systematically increasing

1. concentrations of substances extracted from the earth's crust,
2. concentrations of substances produced by society, and
3. degradation by physical means.

And, in a sustainable society, human needs are met worldwide.

The first three conditions describe an ecologically sustainable society; the fourth ensures that will be socially sustainable.

Green Chemistry

Green chemistry is a philosophy espousing the design of chemical products and processes in ways that reduce or eliminate the use or creation of hazardous substances that pose risks to humans and the environment. Practitioners of green chemistry seek to use safer solvents, run reactions at room temperature, and generally minimize the risk of hazardous chemical releases. Anastas and Warner (1998) developed a set of twelve principles encouraging a more benign, sustainable chemistry, which we will use in this chapter to illustrate the overlaps between sustainability and strategic and financial thinking. Because the principles are clearly delineated and defined in technological, specific terms, they provide an opportunity to link with *existing* financial measures used within in a firm. By incorporating green chemistry principles into investment decisions, managers do not need to adopt new financial tools, for example, triple-bottom-line measures or environmental-activity-based costing (ABC), but can rather use their existing financial tools to make more environmentally informed decisions. To understand why this is the case, it is helpful to be clear on how the principles of green chemistry are applied.

Principles of Green Chemistry

The twelve principles of green chemistry encourage innovation and changes in chemical technology that reduce or eliminate the use of hazardous and toxic substances in the design, manufacture, and use of chemical products (Anastas, Hjeresen, Kirchoff, et al. 2001; Anastas and Kirchoff 2002).

Table 5.1 Twelve principles of green chemistry

1. **Prevention**
   It is better to prevent waste than to treat or clean up waste after it has been created.
2. **Atom Economy**
   Synthetic methods should be designed to maximize the incorporation of all materials used in the process into the final product.
3. **Less Hazardous Chemical Syntheses**
   Wherever practicable, synthetic methods should be designed to use and generate substances that possess little or no toxicity to human health and the environment.
4. **Designing Safer Chemicals**
   Chemical products should be designed to affect their desired function while minimizing their toxicity.
5. **Safer Solvents and Auxiliaries**
   The use of auxiliary substances (e.g., solvents, separation agents) should be made unnecessary wherever possible and innocuous when used.
6. **Design for Energy Efficiency**
   Energy requirements of chemical processes should be recognized for their environmental and economic impacts and should be minimized. If possible, synthetic methods should be conducted at ambient temperature and pressure.
7. **Use of Renewable Feedstocks**
   A raw material or feedstock should be renewable rather than depleting whenever technically and economically practicable.
8. **Reduce Derivatives**
   Unnecessary derivatization (use of blocking groups, protection/deprotection, temporary modification of physical/chemical processes) should be minimized or avoided if possible, because such steps require additional reagents and can generate waste.
9. **Catalysis**
   Catalytic reagents (as selective as possible) are superior to stoichiometric reagents.
10. **Design for Degradation**
    Chemical products should be designed so that at the end of their function they break down into innocuous degradation products and do not persist in the environment.
11. **Real-time Analysis for Pollution Prevention**
    Analytical methodologies need to be further developed to allow for real-time, in-process monitoring and control prior to the formation of hazardous substances.
12. **Inherently Safer Chemistry for Accident Prevention**
    Substances and the form of a substance used in a chemical process should be chosen to minimize the potential for chemical accidents, including releases, explosions, and fires.

Source: (Anastas and Warner 1998), p. 30. Used with permission.

We chose to use the green chemistry framework in our analysis because it was general enough to be widely applicable and specific enough to be interesting, and because it offers a representative microcosm of accepted scientific guidelines for sustainable business practices. Green chemistry implementation also provides data for inclusion in financial equations and analysis.

At first glance, many of the principles may seem foreign and relevant only to professional chemists. But upon further reflection, we find their application to be much broader.

Principle One, "Prevent waste," is familiar to anyone who has worked in corporate environmental sustainability. More than thirty years ago, 3M, a diversified adhesive coatings and adhesive manufacturer, launched its famed PPP ("pollution prevention pays") program, saving the company hundreds of millions of dollars in efficiency improvements while earning numerous awards (Desimone and Popoff 1997).

Principle Two, "Atom economy," refers to the conversion efficiency of chemical reactions. A major goal of green chemistry is to use up all of the reagents—because too often unused reagents become waste. For example, in 2002, Pfizer Pharmaceuticals received the Presidential Green Chemistry Award for its revised manufacturing process of Zoloft, a popular antidepressant drug. The new process doubled product yield, reduced solvent use by tenfold, and eliminated the use of more than 500 metric tons of harsh chemicals (Pfizer Pharmaceuticals 2005). In a very different context the Collins Pine Company found similar success. Collins Pine implemented a program to recover the sander dust produced in producing particle board; by recycling 100 percent of the dust the company is saving an estimated $525,000 per year (Nattrass and Altomare 1999).

Principles Three through Five urge the use of safer syntheses, the creation of safer products, and the elimination of harmful solvents and auxiliary substances. All three recognize the unfortunate fact that many of the chemical reagents and products in widespread use, while effective, are also hazardous to human health and to the natural environment. A well-known example of the benefits of applying these principles involves Rohner Textil AG, a Swiss chemicals manufacturer. After much rethinking, redesign, and testing, Rohner was able to create a new collection of dyes with the desired chemical characteristics. One of the exercise's unanticipated benefits was a huge reduction in the firm's emissions of hazardous chemical waste. Rohner's commitment to the development of nontoxic syntheses and textile dyes was so thorough that when Swiss regulators came to check the quality of the firm's wastewater, they found that water leaving the factory was cleaner than it when it had entered (Werhane, Gorman, and Mehalik 1996a, 1996b).

Principle Six, "Design for energy efficiency," is also a familiar engineering principle, although it often takes a backseat to cost considerations. Many chemical reactions run faster or more efficiently at higher temperatures and pressures, and time is money. Of course, it's also expensive to build, run, and maintain power plants, boilers, and steam generators—and when the total economic and environmental costs are factored in, it may make sense to search for alternative syntheses.

Principle Seven, "Use renewable feedstocks," recognizes that many of Earth's resources are essentially finite and should be conserved for future generations. For example, the petroleum economy, based on solar energy stored more than half a billion years ago, appears to be at its peak. As petroleum supplies become scarcer and more expensive, fuels and feedstocks generated from current solar income (e.g., biofuels, starch-based polymers) will grow in importance and, to the extent that they are manufactured benignly, will contribute to sustainability.

Principle Eight, "Reduce derivatives," refers to the common practice of temporarily masking molecular reactive groups to achieve greater yields. In some cases, reactions simply will not work without blocking groups, and in this instance, green chemists would suggest searching for a different method of synthesis. In other cases, it may be possible to run the reactions, albeit slower. The rationale behind this principle is that as it is often difficult to separate and reuse derivative products, they become waste.

Principle Nine, "Catalysis," expresses a preference for the use of catalysts (compounds reusable after a reaction) over "stoichiometric" reagents. The latter refers to compounds that are actually transformed in a reaction but are not part of the final product. Principles Nine and Two are closely related, as both seek to maximize materials efficiency and minimize waste. For example, the chemical manufacturer Rhône-Poulenc has had substantial success replacing metal-based reagents with zeolite, a more benign alternative (Sheldon 2000).

Principle Ten, "Design for degradation," echoes McDonough and Braungart's contention that "waste equals food." To the extent possible, chemical products should be designed to degrade safely, at the appropriate time, with no persistent toxins. The agricultural giants Archer Daniels Midland and Cargill, both of which have subsidiaries that manufacture biomass-based plastics, exemplify the pursuit of this principle. Their plastic food containers and films decompose naturally and safely under high-temperature and humidity conditions (Dolan 2006).

Principle Eleven, "Real time analysis," involves the adoption of processes and syntheses facilitating real-time monitoring, as opposed to the more traditional approach of sample collection and analysis. Benefits of this analysis include reduced errors, decreased wastes, and quicker response times in the event that something goes wrong.

Principle Twelve, "Inherently safer chemistry," reiterates a focus on "benign chemistry"—expanding the use of water-based solvents, naturally occurring, nontoxic reagents, and room temperature reactions. Approaching the design of chemical processes from this position really opens the door to sustainability, as toxic substances simply don't make it into the equation.

Sustainability practices that improve a company's value proposition are comparable to any action that lowers costs, grows market share and revenues, or enhances brand reputation and, as such, should be a consideration in any cost-benefit analysis. Green chemistry, as will be shown here, also holds potential for liability and risk avoidance.

We propose that these principles are guided by metrics that, when combined with existing finance tools, can improve and create competitive advantage and environmental performance.

Shareholder Wealth and EVA

The generally accepted financial goal for corporate managers in Anglo-American societies is to maximize shareholder wealth, as evidenced in the price of a company's common stock (Brealey and Myers 2005). Shareholders have a residual claim on a firm's income and assets, which means that before they can receive their share, all other claimants (workers, suppliers, debt holders, tax collectors, etc.) must be satisfied first. (In fact, the notion that suppliers of financial capital are entitled to the residual returns generated by a particular business venture is responsible for the "capital" in "capitalism.")

We believe that managers who fail to grasp the logic of environmental sustainability, as represented here by the principles of green chemistry, are not serving their stockholders or stakeholders to their fullest capacity. Ultimately, this may put their firms at a competitive disadvantage or risk charges of fiduciary irresponsibility. Decisions addressing environmental sustainability are neither incongruent nor separate from decisions to maximize shareholder value (Castro and Chousa 2006; White 1996).

Managers essentially have three decisions to make with regard to the funds shareholders have entrusted them with. First, they should attempt to find investment opportunities providing returns above a risk-adjusted hurdle rate (the investment decision). Second, they should attempt to finance these projects at the lowest possible cost (the financing decision). Last, if they are unable to identify any opportunities meeting these criteria, managers should return the shareholders' money (the dividend payout decision).

In making these decisions, managers employ the tools of investment analysis.

One popular tool, EVA, is used to evaluate investment proposals. In its most basic form, EVA measures the amount of income available to a firm's shareholders after subtracting a charge for capital. It incorporates Adam Smith's (1776) notion of *economic profit*, that is, the fundamental observation that businesses must earn a minimum competitive return on *all* sources of capital if they are to remain in existence. Moreover, EVA is popular because it ties operating and financing decisions together into a single performance measure, which can be easily decomposed into measurable, manageable value drivers.

EVA can be expressed in either of two ways (Equations 1 and 2).

(1) EVA = (NOPAT − WACC) × Invested Capital
(2) EVA = (ROIC − WACC) × Invested Capital

where

EVA   = economic value added
NOPAT = net operating profit after tax
            = revenues − operating expense – operating taxes (note that financing costs, e.g., interest expense, are *not* subtracted from revenues to get NOPAT).
WACC = weighted average cost of capital
ROIC   = return on invested capital
            = NOPAT ÷ Invested Capital

In Equation 1, EVA is determined directly from a firm's income statement less a charge for financial capital employed. This charge is determined by multiplying the amount of capital tied up in the project, for example, the manufacturing plant and equipment, by the firm's WACC. The WACC measures the firm's average cost of funds from all financing sources, that is, from both debt and equity holders. Its actual value reflects the risk to which the funds will be exposed and comprises the capital supplier's opportunity cost of capital plus premiums providing compensation for industry, firm, and project risk. Managers can only control firm and project risk. Equation 1 emphasizes the point that financial capital is *not* free; nonproductive assets and wasteful processes generate costs above and beyond material outlays.

Equation 2 is numerically identical to Equation 1, but sometimes provides superior insight. In this equation, EVA is determined by a return spread times the amount of capital employed. A positive spread, where ROIC exceeds the cost of capital, generates positive EVA. Similarly, when an investment is projected to earn less than investors require on their monies, EVA will be negative and the project should not be undertaken.

The Financial Benefits of a Green Chemistry Framework

We now turn to an analysis of the financial and strategic benefits of adopting a "green chemistry" sustainability focus. The results of our efforts, illustrating the expected economic benefits from each of the twelve principles of green chemistry, are summarized in table 5.2.

The first two columns of table 5.2 reflect the expected impact of two key components on a firm's EVA. The third column summarizes the expected impact of each principle on overall EVA (because sometimes there may be conflicting impacts on the EVA drivers). The final column, "Positive Externality?" identifies economic benefits accruing to society at large. Externalities are economic side effects on individuals or entities due to the actions of another individual or entity. These can be negative or positive. Pollution is the classic example of a negative externality; a paper mill's river effluent has a negative economic impact on users downstream, who may need to clean up the water before using it. Pursuit of sustainable business opportunities can also create positive externalities, for example, a reduction of carbon dioxide levels and thus global warming.

This chapter uses the definition of EVA embodied in Equation 1 in its exposition of the financial returns from green chemistry. There are four ways in which the pursuit of green chemistry principles can increase EVA. These are sometimes known as "levers of control" because managers may invoke one or more of them to improve financial performance and establish competitive advantage. To increase EVA, and thus shareholder value, managers might attempt to

1. increase NOPAT by increasing revenues through the sale of "green" products;
2. increase NOPAT by decreasing costs, perhaps by more efficient use of energy;
3. reduce the WACC by reducing present and future risks and communicating this information to capital providers; and
4. decrease the amount of invested capital required—essentially, "do more with less." For example, eliminating toxic solvents reduces the need for air filtration and water purification systems

Table 5.2 Principles of green chemistry and their likely financial effects

Green Chemistry Principle	NOPAT Increase?	WACC Decrease?	Positive EVA?	Positive Externality?
1. Prevent waste rather than treat it after it is formed	**Yes** Cleanup, liability, and insurance cost savings	**Yes** Reduced firm or industry risk factors	**Maybe** If discounted future cleanup costs > cost of preventive design	**Yes** Waste reduction
2. Maximize the incorporation of all process materials into the final product	**Yes** Reduced input and waste disposal costs	**Neutral**	**Yes** Reduced input and waste disposal costs	**Yes** Waste reduction
3. Use and generate substances of little or no toxicity	**Yes** Cleanup, liability, and insurance cost savings	**Yes** Reduced firm or industry risk factors	**Maybe** If discounted future cleanup costs > cost of nontoxic inputs	**Yes** Toxic waste reduction
4. Preserve efficacy of function while reducing toxicity	**Yes** Cleanup, liability, and insurance cost savings	**Yes** Reduced firm or industry risk factors	**Maybe** If discounted future cleanup costs > cost of nontoxic inputs	**Yes** Toxic waste reduction
5. Eliminate or minimize use of or toxicity of auxiliary substances (e.g, solvents)	**Yes** Cleanup, liability, and insurance cost savings	**Yes** Reduced firm or industry risk factors	**Maybe** If discounted future cleanup costs > cost of nontoxic inputs	**Yes** Toxic waste reduction
6. Recognize and minimize energy requirements, shoot for room temperature	**Maybe** If energy cost savings are not offset by new input costs	**Neutral**	**Maybe** If energy cost savings are not offset by new input costs	**Yes** Reduced energy demand

Principle				
7. Use renewable raw material feedstock if economically and technically possible		**Neutral**	**Maybe** If discounted future costs of renewable feedstocks < cost of nonrenewable inputs	**Yes** Toward sustainability
8. Avoid unnecessary derivatization (e.g., blocking group, protection/deprotection)	**Maybe** If input and processing costs do not increase	**Neutral**	**Maybe** If energy cost savings are not offset by new input costs	**Yes** Waste reduction
9. Consider catalytic reagents superior to stoichiometric reagents	**Maybe** If input and processing costs do not increase	**Neutral**	**Maybe** If energy cost savings are not offset by new input costs	**Yes** Waste reduction
10. Design end product to innocuously degrade, not persist	**Maybe** If ultimate disposal is or becomes the responsibility of the manufacturer	**Maybe** If ultimate disposal is or becomes the responsibility of the manufacturer	**Maybe** If ultimate disposal is or becomes the responsibility of the manufacturer	**Yes** Waste reduction
11. Develop analytical methodologies that facilitate real-time monitoring and control	**Maybe** If monitoring costs are offset by savings from a lack of errors, work stoppage and cleanup	**Yes** Reduced firm or industry risk factors	**Maybe** If monitoring costs are offset by savings from a lack of errors, work stoppage and cleanup	**Yes** Fewer disasters and reduced cleanup costs
12. Choose substances/forms that minimize potential for accidents, releases and fires	**Yes** Cleanup, liability, and insurance cost savings	**Yes** Reduced firm or industry risk factors	**Yes** Cleanup, liability, and insurance cost savings	**Yes** Fewer disasters and lower cleanup costs

Source: Anastas and Warner (1998) and author analysis

Revenue Enhancement Strategies

Firms can achieve increased revenues and market share through differentiation and preferred access to markets inaccessible to competitors. Revenue enhancement strategies include, but are not limited to,

1. Sale of Green Products—By meeting "green" consumers' needs, firms can increase sales.
2. Access to Markets—Access to previously, or soon to be, inaccessible markets can be improved.
3. Preferential Purchasing—Current customers may be retained through passing the cost savings generated in production through the supply chain.
4. Increased Innovation—Sustainability thinking can serve as an impetus for innovation; the company that figures out how to comply, and even be in advance of, regulation better, cheaper, and faster will have a competitive advantage (Freeman, Pierce, and Dodd 1999).

Principle Five, "Safer solvents and auxiliaries," led to increased revenues for the consumer products manufacturer SC Johnson. Tasked to reduce the emissions of volatile organic compounds (VOCs) associated with the manufacture of Windex, a popular window and glass cleaner, the firm's chemists redesigned and reformulated the product to use a more benign solvent. The new formula improved the product's cleaning performance by 30 percent and achieved a positive customer response. Between 2002 and 2005, not only did the firm decrease its emissions of VOCs by £1.8 million, but sales increased by 8 percent and market share rose by almost 4 percentage points (Johnson, personal communication, 2005).

Cost Reduction Strategies

Similarly, the ability to produce the same product as competitors at a lower cost is a source of competitive advantage. Bob Willard (2002) has provided a useful framework for considering the costs and benefits of sustainability including, but not limited to,

1. Reduced Operating and Manufacturing Expenses—derived from reuse, waste reduction, and reduced resource consumption (Mehenna and Vernon 2004);
2. Decreased Employee Expense—increased productivity and retention, reduced recruiting expenses because employees feel they work at a "good" company that aligns with their values; and

3. Decreased Insurance Expense—resulting from decreased operating risks.

Principle One, "Prevent waste," and Principle Two, "Maximize incorporation of materials into final product," can again be illustrated by the waste reduction efforts of 3M's PPP program. The company puts its total cost savings at $1 billion and pollution prevention at £2.2 billion (3M 2004). As demonstrated by a split-adjusted stock price increase from $1.96 to $77.86 over the past thirty years, this commitment does not appear to have inhibited 3M's competitiveness; in fact, its CEO, W. James McInerny, cites "a combination of solid top-line growth combined with continued improvement in operational efficiency," which has been driven to a large extent by the PPP program (3M 2004).

Principle Six, "Design for energy efficiency," can clearly have an impact on the net operating profit a business generates, assuming that the cost savings of utilizing alternative energy sources (solar, wind, etc.) are greater than the cost of implementation of new equipment. Although the chemical industry has always pondered problems of energy efficiency, many large companies ranging from UPS to Intel are finding an advantage in reducing energy usage in a variety of industries. Wal-Mart, the world's largest retailer, has set a public goal of reducing energy usage in stores by 30 percent over the next three years in the belief that this movement will garner competitive advantage (Gunther 2006).

Principle Seven, "Use renewable feedstocks," seems relatively straightforward within a chemistry context; it is about reducing the use of petroleum, which is the basis of 98 percent of organic chemicals (Anastas and Kirchoff 2002). If we examine Principle Seven in a broader context, we can also see opportunities for reducing operational expense.

WACC Reduction Strategies

Because investors are assumed to be risk averse—they prefer higher returns to lower returns, and less risk to more risk—any action the company undertakes to reduce firm or project risk should lower the firm's weighted-average cost of capital and thus its hurdle rate for new projects. Sustainability-related strategies that reduce firm risk, and thus WACC, include the following:

1. The reduction or elimination of hazardous chemicals, along with the associated risk of toxic spills, expensive cleanup actions, and adverse publicity
2. Preemptive adoption of higher standards of environmental conduct as a means of avoiding expensive remediation requirements

3. Communicating a firm's sustainability achievements to obtain funding from the socially responsible investment (SRI) community

Principles Three, Four, and Five are all directed at reducing firms' interaction with hazardous chemicals, either in processes or in products. In addition to the expected savings from avoided treatment, transport, storage, monitoring, insurance, and reporting costs, managers might also expect lower financing costs. As a class, chemical manufacturers are likely to pose higher investment risk than, say, food manufacturers or consumer products firms. Johns-Manville (asbestos) and Dow Corning (silicone breast implants) exemplify firms that have gone bankrupt as the result of products later found to be toxic.

Principle Eleven, "Real-time process monitoring," helps firms keep better track of ongoing processes and thus reduce the risk of hazardous spills, which impacts their cost of capital. In fact, communication of reduced risks was a chief impetus for firms adopting the chemical industry's *Responsible Care* principles in the late 1980s (Reinhardt 2000). For example, the 1984 accident at a Union Carbide plant in Bhopal, India, released 40 tons of highly toxic methyl isocyanate and resulted in 15,000 deaths. Investors responded immediately, selling off shares of Union Carbide and other chemical companies as well (Kalra, Henderson, and Raines 1995).

Capital Reallocation Strategies

The goal of capital reallocation strategies is the redistribution of scarce resources (capital) from unprofitable projects to more profitable ones, and sustainability thinking can be extremely helpful in this endeavor. For instance, the focused pursuit of waste reduction, zero emissions, and energy efficiency goals may lead to a reduced need for treatment plants, power plants, and the like. Often these savings are not apparent until a complete "life cycle" accounting of costs is made. Capital reallocation strategies inspired by a commitment to sustainability might include the following:

1. Minimizing the use of virgin materials and maximizing the use of recycling, cogeneration, and take-back systems to avoid expensive preprocessing steps
2. Eliminating the need for hazardous waste treatment and storage facilities
3. Preservation and/or remediation of natural assets that provide ecosystem services

Principle Eight, "Avoid unnecessary derivatization," and Principle Nine, "Consider catalytic reagents superior to stoichiometric reagents," may be the most difficult to apply outside of a chemistry setting. How can we aspire to apply these seemingly technical principles in a business setting? The avoidance of unnecessary derivitization is akin to the avoidance of process steps, similar to the application of lean engineering to streamline processes, thus reducing costs. As demonstrated by the rise of just-in-time delivery systems in virtually every industry, beginning with Toyota in the automotive realm and more recently with Dell in personal computer production, the avoidance of necessary steps not only reduces costs but is invariably associated with increased quality. Another example, outside a business context, is New York City's investment in its Catskills watershed as a means of avoiding construction of an expensive water treatment plant (Chichilnisky and Heal 1998). The law of parsimony applied to most allocation problems can have a positive ecological and economic effect.

Externality Mitigation Strategies

Several of the green chemistry principles are difficult to link with EVA impacts but have a powerful effect on reducing negative externalities. Our contention is that this should garner support at a strategic level, in that strategic decisions are those that impact the firm's success and innovation beyond the limits of a specific project (Green and Hunton-Clarke 2003).

Principal Seven, "Use renewable feedstocks," has no clear impact on EVA, positive or negative; however, by adopting this principle a firm can embrace the idea of reduced oil dependency. As this principle is focused at limiting the use of nonrenewable resources, it enables firms to not only impact the negative external effects their choices have in the realm of consuming limited resources, but also limit emissions of carbon dioxide and other gases that have been linked to global warming. In 2005, $17 billion poured into clean-energy projects in the Unites States—89 percent more than in 2004, as estimated by the research company New Energy Finance Ltd. Worldwide. Clearly the opportunity to become involved in clean energy is being seized by multiple investors; firms can align both innovation and positive externalities by embracing this principle.

Principal Ten, "Design for degradation," impacts multiple externalities. First, by designing to degrade rather than persist, firms can lessen their impact on waste processing at a volume level, by reducing the landfill area required to deal with their products at the end of their useful life cycle. Second, by avoiding the "downcycling" of products as we typically see in efforts to reuse materials like paper and aluminum, design is moved toward a meaningfully sustainable paradigm (McDonough and Braungart 2002). Third, firms can avoid overengineered products that move beyond customers' basic needs and that include elements that present potential medical hazards.

Summary

This chapter clarifies the financial and strategic motivations—what we consider an essential step to making the business case—for sustainability practices by illustrating linkages between green chemistry, strategic competitive advantage, and the measurement of value creation. Our goal is to provide a usable, clear, and concise framework enabling managers and executives to make objective decisions about environmental impacts based on financially and strategically thoughtful logic. Linking the principles of green chemistry with their likely impact on EVA provides managers and engineers with a tool for addressing the costs and benefits of environmental consideration into existing decision processes.

Following this integration, we described the potential strategic levers that could be pulled to utilize this broader set of factors in working to gain a competitive advantage. Through supplying examples of both the direct application of green chemistry principles, and the broader application of the principles outside the chemical industry, we provide insight into the false dichotomy that exists between ecological and economic results. Our hope is that this work will encourage managers and academics to consider the use of green chemistry within the broader context of financial and strategic decisions. With clear principles to follow, and a bit of imagination, there is a path to not only a sustainable environment, but sustainable competitive advantage for organizations bold enough to seek it.

References

3M. 2004. *Annual report.* St. Paul, MN: 3M.

Anastas, P. T., and Kirchoff, M. 2002. Origins, current status, and future challenges of green chemistry. *Accounts of Chemical Research* 35 (9): 686–694.

Anastas, P. T., and Warner, J. C. 1998. *Green chemistry: Theory and practice.* New York: Oxford University.

Anastas, P. T., Hjeresen, D. L., Kirchoff, M., and Ware, S. 2001. Green chemistry: Progress and challenges. *Environmental Science and Technology* 35 (5): 114A–119A.

Brealey, R., and Myers, S. 2005. *Principles of corporate finance.* New York: McGraw-Hill/Irwin.

Castro, N. R., and Chousa, J. P. 2006. An integrated framework for the financial analysis of sustainability. *Business Strategy and the Environment* 15 (5) 322–333.

Chichilnisky, G., and Heal, G. 1998. Economic returns from the biosphere. *Nature* 391 (February): 629–630.

Cloud, P. 1997. Entropy, materials, and posterity. *International Journal of Earth Sciences* 66 (1): 678–696.

Desimone, L. D., and Popoff, F. 1997. *Eco-efficiency: The business link to sustainable development.* Cambridge, MA: MIT Press.

Dolan, K. A. 2006. Revving up nature's engines: Industrial biotechnology is helping replace petroleum in more than gasoline. *Forbes* 178 (2): 76–83.

Freeman, R. E., Pierce, J., and Dodd, R. H. 1999. *Environmentalism and the new logic of business.* New York: Oxford University.

Frosh, R. A., and Gallopoulos, N. 1989. Strategies for manufacturing. *Scientific American* 261 (3): 144–152.

Green, A. O., and Hunton-Clarke, L. 2003. A typology of stakeholder participation for company environmental decision-making. *Business Strategy and the Environment* 12 (5): 292–299.

Gunther, M. 2006. The green machine. *Fortune* 154 (August): 42–57.

Hart, S. L. 2005. *Capitalism at the crossroads.* Upper Saddle River, NJ: Wharton Business School Publishing.

Hawken, P. 1993. *The ecology of commerce.* New York: Collins.

Hawken, P., Lovins, A. B., and Lovins, L. H. 2000. *Natural capitalism: Creating the next industrial revolution.* Boston: Little, Brown.

Hoffman, A. J. 2000. *Competitive environmental strategy.* Washington, DC: Island Press.

Kalra, R., Henderson, G. V., and Raines, G. A. 1995. Contagion effects in the chemical industry following the Bhopal disaster. *Journal of Financial and Strategic Decisions* 8 (2): 1–11.

Larson, A. L. 2000. Sustainable innovation through an entrepreneurship lens. *Business Strategy and the Environment* 9 (5): 304–317.

Larson, A. L., Teisberg, E. O., and Johnson, R. R. 2000. Sustainable business: Opportunity and value creation. *Interfaces* 30 (3):n–12.

McDonough, W., and Braungart, M. 2002. *Cradle to Cradle: Remaking the way we make things.* New York: North Point Press.

McDonough, W., Braungart, M., Anastas, P. T., and Zimmerman, J. B. 2003. Applying the principles of green engineering to Cradle-to-Cradle design. *Environmental Science and Technology* 37 (23): 434A–441A.

Mehenna, Y., and Vernon, P. D. 2004. Environmental accounting: An essential component of business strategy. *Business Strategy and the Environment* 13 (2): 65–77.

Nattrass, B., and Altomare, M. 1999. *The natural step for business.* Gabriola Island, BC: New Society Publishers.

Pfizer Pharmaceuticals. 2005. Environmental initiatives and opportunities: Green chemistry. http://www.pfizer.com/responsibility/ehs/green_chemistry_performance.jsp (accessed August 31, 2006).

Reinhardt, F. 2000. *Down to Earth: Applying business principles to environmental management.* Boston: Harvard Business School Press.

Robèrt, K.-H. 2002a. *Matsushita sustainability report: TVs and refrigerators (internal Matsushita report).* Stockholm, Sweden: The Natural Step International.

Robèrt, K.-H. 2002b. *The natural step story: Seeding a quiet revolution.* Gabriola Island, Canada: New Society.

Robèrt, K.-H., Schmidt-Bleek, F., Aloisi de Lardarel, J., Basile, G., Jansen, J. L., Kuehr, R., et al. 2002. Strategic sustainable development: Selection, design and synergies of applied tools. *Journal of Cleaner Production* 10 (3): 197–214.

Schmidheiny, S. 1992. *Changing course.* Cambridge: MIT Press.

Sheldon, R. A. 2000. Atom efficiency and catalysis in organic synthesis. *Pure and Applied Chemistry* 72 (7): 1233–1246.

Smith, A. 1776. *The wealth of nations.* New York: Oxford University.

Stromberg, M. 2002. William McDonough: Eco-effectiveness. *Professional Builder,* (February). http://www.housingzone.com/probuilder/article/CA462722.html (accessed February 29, 2008).

Werhane, P., Gorman, M. E., and Mehalik, M. M. 1996a. *DesignTex, Incorporated.* Charlottesville, VA: University of Virginia Darden School of Business.

Werhane, P., Gorman, M. E., and Mehalik, M. M. 1996b. *Rohner Textil AG.* Charlottesville, VA: University of Virginia Darden School of Business.

White, M. 1996. Environmental finance: Value and risk in an age of ecology. *Business strategy and the environment* 5 (3): 198–206.

Willard, B. 2002. *The sustainability advantage: Seven business case benefits of a triple bottom line.* Gabriola Island, BC: New Society Publishers.

World Commission on Environment and Development. 1987. *Our common future.* Oxford, UK: Oxford University.

PART III

Exploring New Models for Moving Toward a Sustainable World

CHAPTER 6

Monkey See, Monkey Do? Some Observations on Sustainable Innovations in Zoos

Nicole A. M. Horstman, Frank G. A. de Bakker,
Enno Masurel, and Patricia P. van Hemert

Introduction

The role different types of organizations need to play in the pursuit of sustainable development has received ample attention in the literature, ranging from broad studies on new production and consumption systems and services (Montalvo Corral 2003; O'Brien 1999; Roy 2000) to studies highlighting specific firms and industries (Larson 2000; Little 2005). Innovation has often been included as an important component in such studies (Blättel-Mink 1998; Jorna 2006; Jorna, van Engelen, and Hadders 2004; Ramus 2002; Wheeler and Ng 2004), and the capacity to innovate is regarded as an influential factor in developing and maintaining competitive advantage (Tidd, Bessant, and Pavitt 2001; Tushman 1979). Often, innovation is also linked to entrepreneurship, following Schumpeter's (1934) early suggestion that entrepreneurial innovation is the essence of capitalism and that its associated process of creative destruction is embodied in new products, new production processes, and new forms of organization. In a way, innovation is the key to survival for entrepreneurs, and one could argue that the crucial nature of innovation for their success has made entrepreneurs experts par excellence on innovation. Looking at entrepreneurial behavior, therefore, could deliver important and useful insights on innovation.

It may be for the same reason that solutions to social problems are increasingly sought in the free-market system and business (OECD 2001).

Innovations relating to issues of sustainability are one field in which business is expected to play an important role (cf. Holliday, Schmidheiny, and Watts 2002; Larson 2000). Although a considerable body of research has been published on sustainable entrepreneurship and innovations for global sustainability, this research has overlooked at least one interesting and potentially relevant set of organizations. In this chapter, we focus on sustainable entrepreneurship in zoos: permanent facilities where living animals of wild origin are housed to be exhibited to an audience (Ministerie van LNV 2002). Zoos are confronted with various issues of sustainability and a need to innovate in quite particular ways: they play important roles in a range of biodiversity programs; they inform the general public on issues related to sustainability through nature education and research; and they protect endangered species. As Schaaf (1994, p. 962) notes, "modern zoological institutions are devoting more of their resources to research, education and conservation than they have in the past." Their increasing role in biological conservation programs and research activities is often highlighted (cf. Galbraith and Rapley 2005; Rabb 1994), as are their more traditional educational and recreational functions.

In this chapter we report our observations related to sustainable innovations in zoos in the Netherlands. A brief characterization of these organizations is provided in sidebar 1.

Sidebar 1   Zoos in the Netherlands

The Netherlands is one of the smallest countries in the European Union, with only sixteen million inhabitants. Nevertheless, it has twenty zoos, of which fifteen are members of the Dutch Zoo Federation (NVD). The mission of this association is the protection of nature, spreading nature education, the preservation of endangered species, and the support of wildlife projects. Clustering knowledge and capacities, and thus creating scale economies, is the main reason for NVD's existence. In 2005, almost seven million people visited the five largest Dutch zoos, which makes zoos a significant sector in the Netherlands (Holland.com 2007).

Dutch zoos reach a large part of the Dutch population. People of all ages and with different educational, cultural, social, and ethnical backgrounds are represented in the zoos' audiences. Diverse audiences are typical of zoos and are rarely seen to such an extent in any other kind of organization (NVD 2004). The Dutch zoos compete in a fairly small market for this mixed audience. Meanwhile, they all are mutually dependent, if only for the maintenance of one of their most important assets: their livestock.

These organizations want to distinguish themselves to their audiences, but they need one another as well. These characteristics make zoos an interesting object of research on sustainable innovations. This chapter focuses especially on the way zoos incorporate these innovations in their management practices and on the importance of mutual resource dependence and knowledge exchange between zoos. The central question is, "How can knowledge exchange and resource dependence contribute to the development of sustainable innovations in zoos?" Exploration of this question uncovers several innovative initiatives in zoos that, in turn, provide insights into ways organizations in general can enhance and apply their knowledge base to become more innovative in grappling with issues of sustainability.

The remainder of chapter starts with a brief characterization of (sustainable) innovations and is followed by a discussion of the role of resource dependence and knowledge exchange in facilitating sustainable innovations. That discussion does not seek to address the different theoretical backgrounds of the concepts presented in great detail, but seeks merely to show how these well-known concepts serve as a basis for the empirical research conducted in Dutch zoos. The methods used in the research are then described and the research results presented. The concluding section outlines the lessons learned from the research project and offers five steps to be followed to achieve more sustainable innovations in organizations outside the zoological world.

Innovations and Sustainability

Innovations have been discussed frequently in the literature (cf. Ali 1994; Gopalakrishnan and Damanpour 1997). Five characteristics important for innovations related to sustainability are (1) the risks involved, (2) the various forms in which innovations may appear, (3) the importance of broadening the concept of innovation to include activities that are not purely technology based, (4) the importance and role of knowledge and knowledge sharing, and (5) the role of people in information exchange.

First, innovations are long-term activities that often need significant investments. Attempts at determining their potential success upfront are complicated and often unsuccessful. Depending on the nature of the organization pursuing the innovation and the nature of the innovation, there can be considerable risk for the firm, with more conservative firms likely to avoid these risks more often than do their more entrepreneurial counterparts (Miller and Friesen 1982).

Second, in much of the literature, innovations often seem to be just about materials, machines, and adaptations, limiting the concept to product innovations or process innovations at best. Such a perspective neglects the role of new services, new organizational forms, or new social constructs (Hage 1999)—innovations that are important when addressing issues of sustainability. Hage and Tidd, Bessant, and Pavitt, (2001) see innovations as occurring in dimensions that include people, processes, and products. A growing literature on organizational innovations is also available (cf. Lam 2005) and is increasingly being applied to issues of sustainability (cf. Jorna 2006; Vollenbroek 2002; Wheeler and Ng 2004).

Third, and closely related to the issue of the forms in which innovations can occur, is the tendency to portray only *technological* innovations as "true" innovations. For some, "innovation" means technological improvements (Ten Pas et al. 2002), a definition that highlights only one particular type of innovation. However, innovation can also be seen in a broader way, as indicated by the next characteristic.

Fourth, Nonaka (1994) describes innovation as a process in which an organization creates and defines problems and actively develops new knowledge to solve these problems (Jorna and Waalkens 2004). Innovations then include more than just science and technology; they build on knowledge (Porter and Stern 1999). Organizations develop new products or services by creating and sharing knowledge (Koskinen 2005). Innovation begins with knowledge, expands knowledge, and produces new knowledge (Jorna 2004b).

Finally, knowledge is always linked to people as they carry, produce, and expand knowledge. Innovations are never realized by one single person. Every innovation needs a certain number of people to be successfully developed and implemented. Innovating thus requires teamwork, knowledge and information exchange, combining knowledge from different specialties, etc. Within this process, different forms of knowledge are exchanged to ensure that the idea, product, or service is realized or implemented (Jorna and Waalkens 2004). Entrepreneurial behavior can be an important driver in these exchange processes.

Entrepreneurship

Just as with innovation, there is a vast literature on entrepreneurship (cf. Bögenhold 2004; Schultz 1980; Suarez-Villa 1989). Almost three quarters of a century ago, Schumpeter (1934) stressed the importance of innovative entrepreneurship, marking new firms as drivers of economic growth. It is entrepreneurs who introduce new ideas, new products, and new processes, thereby disrupting current methods of production, organization, and distribution. In a later work, Schumpeter (1943) further states that large firms create barriers to entry for new entrepreneurs because of their access to the accumulated stock of knowledge in specific technological areas; their well-developed competence in large scale R&D projects, production, and distribution; and their access to resources. As such, Schumpeter believes that large firms are more likely to innovate than smaller firms. Whether or not Schumpeter is correct about large firms' inhibiting effect on the emergence of entrepreneurial activity in smaller firms, entrepreneurship remains a phenomenon that takes several forms and appears in small and large firms, in new firms and established firms, in the formal and informal economy, in legal and illegal activities, in innovative and traditional concerns, in high-risk and low-risk undertakings, and in all economic sectors (OECD 1998).

Over time, different authors have stressed different facets of entrepreneurship. Nevertheless, the Schumpeterian concept remains dominant in most of the literature, as witnessed also by the definition of entrepreneurship proposed by the OECD (1998)

> Entrepreneurs are essential agents of change in a market economy, fueling the drive for the increasingly efficient use of resources . . . Growth is promoted when entrepreneurs accelerate the generation, dissemination and application of innovative ideas. Not only do entrepreneurs seek to exploit business opportunities by better allocating resources, they also seek entirely new possibilities for resource use.
>
> (Arzeni 1997, p. 18)

Recently, the network character of entrepreneurship has been gaining attention in the literature (cf. O'Donnell, Gilmore, Cummins, and Carson 2001). In such research, a network stands for a group of firms using combined resources to cooperate on joint projects (Enright 2000; Porter 2000). Entrepreneurs who develop and maintain ties with other entrepreneurs are believed to outperform those who do not. To innovate, entrepreneurs often need to reconfigure relationships with suppliers, and networks can facilitate these processes. Also, networks allow for the sharing of overhead costs and the exploitation of specific scale economies present in collective action (such as bulk purchasing of inputs). On top of that, networks allow for firms to engage in accelerated learning. Networking, as such, seems especially promising for smaller companies as it can pave the way for greater specialization among smaller firms, opening up opportunities for economies of scale and scope. Thus, working together toward sustainable solutions might be one way for zoos, which are smaller organizations, to innovate.

Sustainable Entrepreneurship and Innovations for Sustainability

To bring these perspectives to bear on issues of sustainability, we now turn to innovations that contribute to global sustainability and to the role of sustainable entrepreneurship in creating such innovations. According to Elkington (1999), there were, two decades ago, over one hundred definitions of sustainability and sustainable development. Some of the most interesting challenges are found in between the areas covered by the economic, social, and environmental bottom lines, which Elkington forcefully referred to as "Triple P": profit, people, and planet. Applying ideas on sustainability to entrepreneurship and innovation leads to the notions of sustainable entrepreneurship and sustainable innovations. Sustainable entrepreneurship may be defined as "leading the firm in making balanced choices between profit, people and planet, without compromising future generations" (Masurel 2007, p. 191). Sustainable entrepreneurship means that a company's processes are organized in such a way that the company not only uses the environment of which it is part, but that it also interacts with that environment in a nonharmful way and hopefully even contributes to, or enhances, the health of that environment. Such a relationship means that the environment does not suffer from entrepreneurial activities and that innovations and can potentially benefit from them (Jorna 2004a). Sustainable entrepreneurship also involves adapting and renewing existing knowledge and creating new knowledge. The process of innovation aims to create sustainable and preservable knowledge.

Developing sustainable innovations implies that an organization changes its behavior to move in a more sustainable direction. In other words, the organization changes the balance between profit, people, and planet. Sustainable entrepreneurship cannot be discussed without mentioning innovation, because it has much to do with adopting new production technologies and organizational processes (cf. Holliday, Schmidheiny, and Watts 2002; Zwetsloot 2001). Rennings (2000) speaks of the technological, social, and institutional innovations that are at play in discussing sustainable innovation. Sustainable development hence must entail a process of change in which the exploitation of resources, the direction of investments, the orientation of technological developments, and institutional change are all made consistent with future as well as present needs. Exchange of knowledge and resources among different organizations is frequently helpful in efforts to evoke such a process of change.

Knowledge Exchange and Resource Dependence

In essence, innovation deals with the knowledge-related processes of an organization. It influences the way an organization handles knowledge. After all, when innovating, either in a conventional or in a sustainable way, changes take place in the organizational-knowledge processes (Jorna and

Waalkens 2004). When organizations want to create sustainable innovations that are successful, they will have to adapt; the "sustainability feeling" slowly has to be fitted into the organization's strategy (Little 2005). By adapting, new knowledge is created and added to the existing knowledge base, which is changing constantly (Jorna 2004a). Thus, when organizations want to work together in the field of (sustainable) innovations, they have to exchange knowledge, because knowledge forms the base for developing (sustainable) innovations.

As indicated earlier, networks can play an important role in establishing such sustainable innovations, because knowledge can easily be exchanged and adapted within groups. More and more organizations are engaged in knowledge exchange or innovation networks to use knowledge and technology as efficiently and effectively as possible (Ten Pas et al. 2002). Research has pointed out that interaction between organizations and networks is an important source for new knowledge and innovation (Dougherty 1992; Goes and Park 1997; Nohria and Eccles 1992) By making use of a network, organizations can exchange knowledge, expertise, and other resources that they need to stimulate innovation (Alter and Hage 1993; Swan and Newell 1995).

But networking as a business strategy requires investments in social communication, informal bonds, training, and education. To build up and operate networks effectively, time and effort are needed. Furthermore, networking may be a desirable or necessary condition, but it is by no means sufficient to ensure good entrepreneurship. And it may even contradict the entrepreneurial spirit by stimulating uniformity.

To understand the relevance of knowledge exchange, it is useful to look at the notion of resource dependence. Resource-dependence theory has primarily been developed by Pfeffer and Salancik (1978). According to them, an organization's possibility to gain and maintain resources is key to organizational survival. Goes and Park (1997) state that every organization depends on its environment for scarce and valuable resources to perform certain activities. This statement follows from the assumption that organizations are open systems and therefore (1) are not self-supporting, (2) cannot produce all needed resources internally, and (3) have to mobilize resources from other organizations in their environment to survive (Goes and Park 1997; Meeus, Oerlemans, and Hage 2001). To gain access to the required resources, interaction on a regular basis with other organizations that possess these resources is needed (Pfeffer and Salancik 1978).

The mutual dependence among different organizations is the factor that could stimulate them to cooperate. Two main strategies are often distinguished in dealing with situations of resource dependence: buffering and bridging. An organization either tries to establish a buffer to prevent situations of high interdependence with its environment or tries to bridge these dependencies through some form of cooperation with other organizations that control the resources it requires (Meznar and Nigh 1995). Bridging thus requires adaptation, as Meznar and Nigh argue. Yet, it allows the formation of networks and could allow smaller firms to engage in innovative entrepreneurial activities, for instance, on issues of sustainability.

Taking these theoretical perspectives together enables us to study the way a specific set of organizations is trying to meet the changing demands related to sustainability, using one another's knowledge to improve both individual and group performance, while competing with each other as well. Before turning to the methods used to study the sustainable initiatives and practices of Dutch zoos and the results of the study, the empirical context of the research—the nature and situations of Dutch zoos—is presented.

Sketching Out the Empirical Context: Dutch Zoos

Zoos are a special type of organization in which issues of sustainability are highly important in several ways. Historically, zoological gardens fulfilled mainly recreational and educational functions as "living museums" and menageries, respectively (Rabb 1994). With their increasing focus on research and conservation, their role has changed over time. As Rabb (Rabb 1994, p. 159) indicates, "Education is the primary function in conservation, but zoos have begun to make significant contributions as genetic refuges and reservoirs, especially for large vertebrate species threatened with extinction." Zoos thus play an important role in biodiversity programs and in informing the general public on nature education, research, and the protection of endangered species. Animals are an important element of the "planet" aspect in the Triple-P notion; animal well-being is an important part of a commitment to a just world, and biodiversity loss is a danger to all species, especially to humans. Although zoos are one of the most appealing organizations in which animals play a key role, until today hardly any attention has been paid to the role zoos play in the field of sustainable development, let alone what other organizations can learn from them.

The mission statement of the NVD pledges "to support her members in optimally informing a big audience about the living nature and nature protection." The main goal of this mission is to create awareness and respect for the importance of nature. This awareness has to lead to a change of attitude toward nature and its preservation, as well as to a responsible use of natural resources (NVD 2004, p. 11). Thus, zoos are closely involved with the pursuit of a sustainable world and are well aware of their specific position on this issue.

To fulfill their tasks, zoos work together and exchange knowledge on a regular basis—for example, about the development of breeding programs in order to preserve endangered species (Zodiac Zoos 2006). One of the reasons that zoos exchange a large amount of knowledge is their high mutual interdependence, stemming from the fact that there are not many other sources for the often specialized knowledge they need. Zoos therefore rely on their direct competitors, nationally and internationally, for information on animal housing, animal feeding, medical attendance, and breeding programs. Together with hundreds of other zoos, Dutch zoos are part of an international network that contributes to raising awareness of the importance of nature conservation. Hence, zoos regularly cooperate and exchange knowledge concerning sustainable innovations—for example, the development of species protection programs. Zoos also experiment with water-treatment methods, alternative-exhibition methods, and energy-conservation programs (Avifauna 2006; Dierenpark Emmen 2006; Zodiac Zoos 2006). As Rabb notes, "North American, European, and Australian zoos are meanwhile assisting the development of technical capacities among zoo counterparts, government agencies, and protected areas in both developing and developed countries of the world to further the conservation of biodiversity" (Rabb 1994, p. 159).

These organizations have mainly been neglected in the literature on sustainable entrepreneurship and sustainable innovations, but they do provide an interesting angle to this literature because of their strong resource dependence that has led them to specific forms of knowledge exchange. This knowledge exchange potentially enhances the development of sustainable innovations. Knowledge exchange can stimulate innovation, because by combining existing knowledge, new knowledge can be created, which asserts positive effects on the development of innovations. Studying zoos, therefore, can teach us how the development of sustainable innovations can be facilitated in other types of organizations. The next section describes the methodological choices that were made to conduct this study.

Methods

A series of interviews were conducted in Dutch zoos to study how knowledge exchange and resource dependence could contribute to the development of sustainable innovations. Managers at eight different zoos, seven of which were NVD members, were interviewed. The Dutch government recognizes twenty zoos as "real" zoos based on factors such as the number of species housed. Therefore, these interviewees represented 40 percent of the relevant population of zoos in the Netherlands and can be considered to be quite representative of the entire population.

In the interviews, respondents were asked to reflect on innovation in general, sustainable innovation, knowledge exchange, resource dependence, and competition. All interviews typically lasted about an hour, were recorded and transcribed, and were conducted using a semistructured interview scheme. This approach allowed respondents to digress from the original questions asked and to add valuable insights. All the interview transcripts were combined and some data matrices were drawn to provide an overview of the findings on some salient issues. Additional data were gathered through published reports and websites.

Sustainable Innovations in Zoos

Innovations in zoos can be seen in the context of creation or development of a product, process, service, or technology through which change, renewal, and better performance or other advantages are created. Important in this process is the improvement of the well-being of animals. When innovating, zoos do not only look at advantages for the organization but also look at the effects these innovations will have on the animals.

From interviews it appears that innovations in zoos are, almost by nature, always sustainable innovations. The most important reason why zoos practice sustainability is that they have a certain social responsibility with regard to their natural environment, which compels them to focus on sustainability. Sustainable innovations in zoos can be defined as innovations that evolve around people and the planet (especially animals) instead of merely around financial or economic performance. By definition, sustainable innovations also respect current and future generations.

The sustainable innovations reported in the interviews focused on the health of the livestock, the use of solar energy, water purification, and reuse of all kinds of materials. Some specific examples mentioned in the interviews include:

- new medical programs to cure animals;
- waste reduction through the use of bio-floors;
- vermin control through housing meerkats with apes;
- reuse of organic waste;
- sand filtering;
- use of springing floors for elephants (in order to protect their feet); and
- recycling of building material

Sustainable innovations like the ones listed above are usually developed through internal propositions. They sometimes are reported to stem from brainstorming sessions and often develop through trail and error. However, external influences sometimes also stimulate sustainable innovations. Zoos, for instance, can get in touch with organizations in their business environment that stimulate them to develop alternative approaches, or, and this aspect was emphasized in the interviews, they can collaborate with other zoos. Knowledge exchange plays an important role in these situations.

Zoos exchange knowledge with government agencies, universities, and all kinds of other organizations, but fellow zoos are the most important parties in their network. With them, they communicate mostly in an informal way. Subjects of knowledge exchange between the zoos are breeding programs, animal feeding, the improvement of animal transport and animal housing, specific knowledge about species, and education of visitors. This exchange takes place between different groups and at all levels: CEOs, technicians, vets, caretakers. The process of knowledge exchange is normally give-and-take: when a zoo asks for certain information, it is likely to be provided by another zoo, and vice versa. Knowledge exchange, therefore, is a mutual thing: it only works if it comes from both sides. As one respondent noted:

> In principle, knowledge on innovations is exchanged with everybody. Yet, it depends from zoo to zoo; with certain zoos we have a better relationship and we're able to communicate easier than with others.

Knowledge exchange positively influences the development of sustainable innovations, not only by creating new ideas, but also by combining existing ideas. Knowledge exchange and the development of sustainable innovations reinforce one another. An organization initiates a knowledge exchange to create improvements and, in this initial sharing, additional knowledge is often shared in return; then, later, information about the improvement is shared as it yields useful results. Innovating and exchanging knowledge about innovations is a continuous process from which an organization can benefit. This exchange of knowledge can be stimulated through a situation of resource dependence. The network between zoos plays an important role in developing sustainable innovations. Although zoos can be seen as each other's competitors in a certain sense, they also depend heavily on each other for knowledge. As one respondent noted:

> If things go wrong in Dutch zoos, the audience will remember; this will affect our zoo as well, our image.

It is impossible for a zoo to have all the necessary information it needs at its disposal, and therefore all zoos depend in one way or another on other zoos, both for information and for animals to maintain their own collection. Breeding programs are a mutual interest of zoos both to maintain the species of animals and to prevent overpopulation. Knowledge exchange and resource dependence are of great importance to the sustainable and innovative behavior of zoos. They depend on each other for the well-being of their animals, their main asset. This is a key reason why zoos are willing to help one another: when animals in one zoo are taken care of in the right way, this good treatment can be in the best interests of all zoos.

> Knowledge exchange has to be reciprocal. If one zoo never gives any information to us, we're not likely to share information either.

In these organizations, sustainable innovations are driven by the specific characteristics of zoos. Nevertheless, several respondents also indicated reasons not to engage in these innovations—cost and associated risks being the most prominent ones. The responses of those interviewees suggest that the discussion of entrepreneurial behavior and the differences between more conservative and more entrepreneurial organizations might well fit the differences between more conservative zoos and more entrepreneurial zoos.

Concluding Observations

In this chapter the concepts of sustainable innovations, resource dependence, and knowledge exchange are combined. Zoos, like other organizations, cannot be expected to have control over all the information they need. Different organizations have different ways of compensating for this lack of knowledge. Because of their mutual dependence, the zoos in this study depend heavily on one another to acquire the knowledge they need to innovate. Other organizations may also choose to do so or might need to do so. This situation is depicted in figure 6.1.

The central question of this chapter was, "How can knowledge exchange and resource dependence contribute to the development of sustainable innovations in zoos?" The interviewees offered a number of examples and shared some further insights on sustainable innovation processes in situations of

Figure 6.1 Resource dependence and the continuous interaction between knowledge exchange and innovation

resource dependence. Of course, it is known that resources can be transformed into innovations (Meeus, Oerlemans, and Hage 2001). It is especially clear that exchanging and combining knowledge positively influences the development of (sustainable) innovations. This observation is consistent with the statements of Håkansson (1990) and Damanpour (1996) that interaction with organizations that possess knowledge in other areas can produce new ideas and therefore new knowledge. In consequence, a bigger knowledge base enhances the development of new ideas and innovations. Innovating is also easier when knowledge is gained from other organizations, because then the knowledge is available immediately and does not have to be developed from scratch (Cavusgil, Calantone, and Zhao 2003). Furthermore, complex innovation processes cannot always be carried out by individual organizations (Arias and Fischer 2000; Koskinen 2005). In these circumstances, knowledge exchange can be a useful solution to organizations.

Up until this point the conclusions of this research are in accordance with the literature consulted. The insight these interviews produced is not only that knowledge exchange stimulates the development of (sustainable) innovations but that the development of (sustainable) innovations also stimulates knowledge exchange. This reciprocal relationship means that there is a continuing process that positively influences the development of (sustainable) innovations: the exchange of knowledge stimulates the development of new ideas over which new knowledge can be exchanged, etc. Figure 6.1 outlines this situation. This perspective implies that organizations exchange resources not only because they have no other choice, but also because it is to their own advantage. This reciprocal relationship is also demonstrated by the observation that knowledge sharing has to be mutual: "You scratch my back and I'll scratch yours." When organizations share knowledge, they expect something in return, something they can use to their own advantage.

What is important in knowledge exchange is the degree of resource dependence—when organizations are interdependent they exchange more knowledge than when they are not. This seems like an obvious conclusion, but it is not explicitly mentioned in any literature and has not been tested empirically before. Oerlemans, Meeus, and Boekema (1998) do indicate that the resource dependence theory is a much-used theory to explain collaboration between organizations, because organizations need one another for certain resources and therefore exchange them among themselves. It would be interesting to investigate, based on our empirical findings, the possibility that the higher the interdependence between zoos, the more knowledge they exchange.

Naturally many questions cannot be answered from these interviews, but the interviews can raise interesting ones for future exploration.

Two important questions that flow naturally from the experiences of these zoos are, (1) does knowledge exchange influence resource dependence? and (2) does resource dependence directly influence innovation?

The first question deals with the issue of whether or not organizations become more interdependent when they exchange more knowledge The results of this research do not show whether or not this is the case; such a possible relationship was not explicitly tested in this exploratory study. To answer this question, further research has to be conducted. It could be interesting to know whether organizations become more interdependent with other organizations when they exchange more knowledge, especially because most organizations strive to be as independent as possible. The answer to this question can be explored by doing research in different organizations to see how much knowledge they exchange with other organizations, and under what conditions. This could be done by means of a survey or network analysis (Hargie and Tourish 2000). Network analyses can be used to measure information exchange and can therefore be useful in measuring knowledge exchange. In these analyses, the kind of knowledge and the frequency of knowledge exchange can also be measured.

It would also be interesting to examine whether organizations that exchange knowledge on a large scale are becoming more dependent on one another. Using a questionnaire one could try to determine an organization's (reported or actual) dependence. This information should help in gaining insight into whether or not resource dependence can influence innovation without knowledge exchange being a mediator.

Finally it would be interesting to learn to what extent the results of this research could be generalized for other organizations. For instance, it is possible that the degree of resource dependence and, with that, the degree of knowledge exchange among museums, is comparable with that among zoos, because zoos are museums in a way. Similar patterns with regard to innovation and knowledge exchange might be found in museums and similar organizations. Other organizations in which (sustainable) innovations could be stimulated through knowledge exchange and resource dependence are specialized organizations in relatively small lines of business, without much competitiveness, because the organizations all have their own specialty. This perspective once again brings up the ideas discussed on entrepreneurship and particularly Schumpeter's (1943) emphasis on the important role of large firms. As discussed earlier, to achieve innovation, entrepreneurs reconfigure relationships within their networks. Working through their networks, voluntarily or more-or-less forced because of a situation of high resource dependence, firms will strive for new opportunities. The examples reported on sustainable solutions sought in zoos provide some ideas of how these processes are configured. Therefore, this chapter concludes with five steps that may contribute to a greater number of sustainable innovations.

Sidebar 2   Developing sustainable innovations in five steps

1. Identify which competitive organizations have valuable information for improving your organization's sustainability performance.
2. Identify what kind of sustainability information within your organization may be of interest to the organizations you identified in Step 1.
3. Get in touch with these organizations and try to formulate terms in which information may be exchanged.
4. Exchange the information and seek for a long-time balance between the information put in and that pulled out.
5. Keep on looking for new partners.

References

Ali, A. 1994. Pioneering versus incremental innovation: Review and research propositions. *Journal of Product Innovation Management* 11 (1): 46–61.

Alter, C., and Hage, J. 1993. *Organizations working together.* Newbury Park, CA: Sage.

Arias, E. G., and Fischer, G. 2000. *Boundary objects: Their role in articulating the task at hand and making information relevant to it.* Paper presented at the International ICSC Symposium on Interactive & Collaborative Computing (ICC'2000), University of Wollongong, Australia.

Arzeni, S. 1997. Entrepreneurship and job creation. *OECD Observer* (209): 18–20.

Avifauna. 2006. Fokprogramma's. http://www.avifauna.nl (accessed April 14, 2006).

Blättel-Mink, B. 1998. Innovation towards sustainable economy: The integration of economy and ecology in companies. *Sustainable Development* 6 (2): 49–58.

Bögenhold, D. 2004. Entrepreneurship: Multiple meanings and consequences. *International Journal of Entrepreneurship and Innovation Management* 4 (1): 3–10.

Cavusgil, T., Calantone, R., and Zhao, Y. 2003. Tacit knowledge transfer and firm innovation capability. *Journal of Business & Industrial Marketing* 18 (1): 6–21.

Damanpour, F. 1996. Organizational complexity and innovation: Developing and testing multiple contingency models. *Management Science* 42 (5): 693–716.

Dierenpark Emmen. 2006. Milieubarometer. http://www.dierenpark-emmen.nl (accessed March 1, 2006).

Dougherty, D. 1992. Interpretative barriers to successful product innovation in large firms. *Organization Science* 3 (2): 179–202.

Elkington, J. 1999. *Cannibals with forks.* Oxford: Capstone.

Enright, M. J. 2000. The globalisation of competition: Policies towards regional clustering. In *The globalisation of multinational enterprise activity and economic development*, ed. N. Hood and S. Young, pp. 303–331. London: MacMillan.

Galbraith, D., and Rapley, W. 2005. Research at Canadian zoos and botanical gardens. *Museum Management and Curatorship* 20 (4): 313–331.

Goes, J., and Park, S. H. 1997. Interorganizational links and innovation: The case of hospital services. *Academy of Management Journal* 40 (3): 673–696.

Gopalakrishnan, S., and Damanpour, F. 1997. A review of innovation research in economics, sociology and technology management. *Omega* 25 (1): 15–28.

Hage, J. 1999. Organizational innovation and organizational change. *Annual Review of Sociology* 25: 597–622.

Håkansson, H. 1990. Technological collaboration in industrial networks. *European Management Journal* 8 (3): 371–379.

Hargie, O., and Tourish, D. 2000. Data collection log-sheet methods. In *Handbook of communication audits for organisations*, ed. O. Hargie and D. Tourish, pp. 104–127. London: Routledge.

Holland.com. (2007). Top 20 bezoekers Nederlandse attractieparken, dierentuinen en musea. http://www.holland.com/files/corporate/onderzoek/attracties_%202005.pdf (accessed October 21, 2007).

Holliday, C., Schmidheiny, S., and Watts, P. 2002. *Walking the talk: The business case for sustainable development*. San Francisco: Greenleaf.

Jorna, R. 2006. Knowledge creation for sustainable innovation: the KCSI programme. In *Sustainable Innovation: The organisational, human and knowledge dimension*, ed. R. Jorna, pp. 2–14. Sheffield: Greenleaf.

Jorna, R. 2004a. Duurzaamheid: van mileu en techniek naar menskunde en organisaties [Sustainability: from environment and technology to people and organization]. In *Duurzame innovatie. Organisaties en de dynamiek van kenniscreatie*, ed. R. Jorna, J. van Engelen and H. Hadders, pp. 45–60. Assen: Koninklijke Van Gorcum.

Jorna, R. 2004b. Kennis als basis voor innovatie: management en creatie [Knowledge as a basis for innovation: management and creation]. In *Duurzame innovatie. Organisaties en de dynamiek van kenniscreatie*, ed. R. Jorna, J. van Engelen and H. Hadders, pp. 70–95. Assen: Koninklijke Van Gorcum.

Jorna, R., and Waalkens, J. 2004. Innovatie: niet ondubbelzinnig, wel belangrijk [Innovation, many headed and certainly important]. In *Duurzame innovatie. Organisaties en de dynamiek van kenniscreatie*, ed. R. Jorna, pp. 31–44. Assen: Koninklijke Van Gorcum.

Koskinen, K. 2005. Metaphoric boundary objects as co-ordinating mechanisms in the knowledge sharing of innovation processes. *European Journal of Innovation Management* 8 (3): 323–335.

Lam, A. 2005. Organizational innovation. In *The Oxford handbook of innovation*, ed. J. Fagerberg, D. C. Mowery and R. R. Nelson, pp. 115–147. Oxford: Oxford University.

Larson, A. L. 2000. Sustainable innovation through an entrepreneurship lens. *Business Strategy and the Environment* 9 (5): 304–317.

Little, A. D. 2005. How leading companies are using sustainability-driven innovation to win tomorrow's customers. http://www.adl.com (accessed May 17, 2006).

Masurel, E. 2007. Why SMEs invest in environmental measures: Sustainability evidence from small and medium-sized printing firms. *Business Strategy and the Environment* 16 (3): 190–201.

Meeus, M., Oerlemans, L., and Hage, J. 2001. Patterns of interactive learning in a high-tech region. *Organization Studies* 22 (1): 145–172.

Meznar, M., and Nigh, D. 1995. Buffer or Bridge? Environmental and organizational determinants of public affairs activities in American firms. *Academy of Management Journal* 38 (4): 975–996.

Miller, D., and Friesen, P. 1982. Innovation in conservative and entrepreneurial firms: Two models of strategic momentum. *Strategic Management Journal* 3 (1): 1–25.

Ministerie van LNV. 2002. Dierentuinenbesluit. from http://wetten.overheid.nl/cgi-bin/sessioned/browsercheck/continuation=25967-002/session=046321607004340/action=javascript-result/javascript=yes (accessed February 2, 2007).

Montalvo Corral, C. 2003. Sustainable production and consumption systems—cooperation for change: Assessing and simulating the willingness of the firm to adopt/develop cleaner technologies. The case of the In-Bond industry in northern Mexico. *Journal of Cleaner Production* 11 (4), 411–426.

Nohria, N., and Eccles, R. G. 1992. *Networks and organizations*. Boston: Harvard Business School.

Nonaka, I. 1994. A dynamic theory of organizational knowledge creation. *Organization Science* 6 (1) 14–37.

NVD: Nederlandse Vereniging van Dierentuinen. 2004. *Visie op verder*. Amsterdam: Nederlandse Vereniging van Dierentuinen.

O'Brien, C. 1999. Sustainable production: A new paradigm for a new millennium. *International Journal of Production Economics* 60–61: 1–7.

O'Donnell, A., Gilmore, A., Cummins, D., and Carson, D. 2001. The network construct in entrepreneurship research: A review and critique. *Management Decision* 39 (9): 749–760.

OECD. 1998. *Fostering entrepreneurship*. Paris: Organisation for Economic Cooperation and Development.

OECD. 2001. *Corporate social responsibility. Partners for progress*. Paris: Organisation for Economic Cooperation and Development.

Oerlemans, L., Meeus, M., and Boekema, F. 1998. Do networks matter for innovation? The usefulness of the economic network approach in analysing innovation. *Tijdschrift voor Economische en Sociale Geografie* 89 (3): 298–309.

Pfeffer, J., and Salancik, G. 1978. *The external control of organizations: A resource dependence perspective*. New York: Harper & Row.

Porter, M. E. 2000. Location, competition and economic development: local clusters in a global economy. *Economic Development Quarterly* 14 (1): 15–34.

Porter, M. E., and Stern, S. 1999. *The new challenge to America's prosperity: Findings from the innovation index*. Washington: Council on Competitiveness.

Rabb, G. 1994. The changing roles of zoological parks in conserving biological diversity. *Integrative and Comparative Biology* 34 (1): 159–164.

Ramus, C. 2002. Encouraging innovative environmental actions: What companies and managers must do. *Journal of World Business* 37 (2): 151–164.

Rennings, K. 2000. Redefining innovation: Eco-innovation research and the contribution from ecological economics. *Ecological Economics* 32 (2): 319–332.

Roy, R. 2000. Sustainable product-service systems. *Futures* 32 (3–4): 289–299.

Schaaf, D. 1994. The role of zoological parks in biodiversity conservation in the Gulf of Guinea islands. *Biodiversity and Conservation* 3 (9): 962–968.

Schultz, T. W. 1980. Investment in entrepreneurial ability. *Scandinavian Journal of Economics* 82 (4): 437–448.

Schumpeter, J. 1934. *The theory of economic development.* Cambridge: Harvard University.

Schumpeter, J. 1943. *Capitalism, socialism and democracy.* London: Allen and Unwin.

Suarez-Villa, L. 1989. *The evolution of regional economies: Entrepreneurship and macroeconomic change.* New York: Praeger.

Swan, J., and Newell, S. 1995. The role of professional associations in technology diffusion. *Organization Studies* 16 (5): 847–874.

Ten Pas, I., Goedegebuure, L., Huisman, J., and Jongbloed, B. 2002. *Kennis maken in de regio. Een verkennend onderzoek naar kennistransfer en kennisrelaties.* Enschede: Center For Higher Education Policy Studies.

Tidd, J., Bessant, J., and Pavitt, K. 2001. *Managing innovation: Integrating technological, market and organisational change.* Bognor Regis: Wiley.

Tushman, M. 1979. Managing communication networks in R&D laboratories. *Sloan Management Review* 20 (2): 37–49.

Vollenbroek, F. 2002. Sustainable development and the challenge of innovation. *Journal of Cleaner Production* 10 (3): 215–223.

Wheeler, D., and Ng, M. 2004. Organizational innovation as an opportunity for sustainable enterprise: Standardization as a potential constraint. In S. Sharma and M. Starik (Eds.), *Stakeholders, the environment and society,* pp. 185–211. Sheffield: Greenleaf.

Zodiac Zoos. 2006. Toekomstplannen Zodiac Zoos. http://www.zodiaczoos.nl/index.asp (accessed April 14, 2006).

Zwetsloot, G. 2001. The management of innovation by frontrunner companies in environmental management and health and safety. *Environmental Management and Health* 12 (2): 207–214.

CHAPTER 7

Toward Environmental Sustainability: Developing Thinking and Acting Capacity within the Oil and Gas Industry

Laurie P. Milton and James A. F. Stoner

"With few exceptions, top managers feel responsible to create organizations well prepared to thrive in the future. This has become an especially relevant goal, as the future more rapidly becomes the present. As a consequence . . . more so than ever before managers are actively seeking ideas and insights for creating firms that will succeed in the business environments that lie ahead" (Huber 2004). It can easily be argued that firms best suited to the business environment of tomorrow will be those that take environmental sustainability seriously and that are and are seen to be environmentally sustainable themselves. Leaders everywhere are struggling to figure out what this means for their industry and for their companies.

Clearly, the public at large and the scientific community are concerned about Earth's capacity to support life. Scientific arguments suggest that the ways we acquire and use energy, air, and water emissions, overuse water, leave uncontrolled development footprints and deforest tracts of land are major contributors to the environmental degradation we are experiencing.

The acquisition and use of energy are particularly serious concerns at present because of the environmental, political, and economic costs of acquiring energy and the environmental damage that often results from our use of it. Reducing emissions, harvesting energy in environmentally sensitive ways, and decreasing the demand for fossil fuel energy are all important changes to be made. Each is necessary. As long as there is a demand for fossil fuel energy, the oil and gas industry will be called upon to supply it within socially and legally acceptable limits. Industry is in a tough spot—damned if they do and damned if they don't. There are 6.6 billion people on this planet. Each and every one of them requires energy in some form to survive. Basic energy needs will likely increase as near 80 million people are born each year (Central Intelligence Agency 2007). Economic growth, estimated to be near 5 percent in 2007, is further driving the demand for energy. Yet, societies seem to be of multiple minds as they hold industry accountable for environmental degradation, fight for a share of resource revenues, and continue to consume energy.

To create a sustainable world, society and industry must work together to produce and use energy in environmentally sustainable ways. Individuals, families, communities, and businesses could consume less. They and the financiers who serve them could invest in firms that are environmentally responsible. Industry could harvest energy in environmentally sustainable ways. Companies could cooperate with one another to create and implement environmentally sustainable business practices. Society could set up a structure that supports all of this (e.g., underwriting technology development, legislating environmental compliance, and constraining the ability of companies to harvest energy in countries that have low environmental standards). Politicians could use their power of persuasion and public policy to advance sustainability. Consumers could pay more for the energy they use and thereby both share the costs of an environmentally sustainable industry and encourage themselves to use less energy. Each party could follow these and other courses of action to create a sustainable world. Each *could* do so because it would be the right thing to do? But, will they? What would it take to get them to want to create a sustainable world? To get them to act in this direction, as clearly they must do?

Perhaps the most important change that is required is a new level of thinking—a willingness to think about doing the impossible and a willingness to commit to making the impossible possible. As Albert Einstein is often quoted to have said: "The problems that exist in the world today cannot be solved at the same level of thinking that created them."

The world is interdependent and we (meaning all of us) must learn to think and act interdependently—to cooperate in areas where we used to compete. Finger pointing and demonizing one another will undermine our capacity to do so.

Industry and society must *meaningfully* integrate environmental, social, and economic sustainability into their mental models of business, government, and daily life. And they must integrate this thinking of sustainability into their decision-making processes. Both must optimize their thinking capacity to develop, integrate, and retain knowledge about sustainability issues and solutions. All must develop the willpower, puzzle-solving, and execution capabilities required to address cooperatively issues they may individually feel are daunting.

This chapter focuses on industry's role in creating an environmentally sustainable world. Our aim is to introduce a base set of competencies and actions that research suggests companies develop to succeed in fast-paced, changing (especially threatening) contexts where expertise is distributed across people and organizations and outcomes are critical. These base competencies provide a solid (and, we believe, necessary) foundation that will enable companies to succeed in becoming environmentally sustainable.

Although our argument applies to other industries, we focus on just one industry—the global oil and gas industry—and examine just one part of that industry—the Canadian oil and gas industry—in order to illustrate our science-based perspective and consider, by way of example, what the oil and gas industry, and other industries, can do to create and thrive within an environmentally sustainable world. The choice of industry is fortuitous for this inquiry because it is an industry that is important, complex, and rich in many emotionally charged historical and political events. It is also an industry that must undergo a major transformation if its corporate members are to survive as ongoing business entities. After all, existing sources of fossil fuel may eventually be exhausted. And, if we do not transform the ways in which we acquire and utilize our energy resources, our ways of acquiring and using them may damage the physical environment so severely that society as we know it will cease to exist well before even the currently known and proven fossil fuel resources are exhausted. Some argue that our planet will become incapable of supporting human life. Others suggest that the population will be thinned out and have to survive in a much less friendly environment in which it uses what still works of our current physical and social infrastructure as best it can. Either and all points between are scary enough to warrant prudence. There are many reasons to believe that industry can simultaneously address environmental concerns and continue to thrive.

The Canadian oil and gas industry is a major industry in terms of economics and social, cultural, and environmental impacts, even though it is a modest part of the total global industry, producing a mere 3.9 percent of the world's petroleum in 2006 (BP p.l.c. 2007). It also has many of the same opportunities, problems, constraints, and resources found in other segments of the global oil and gas industry and in many other, entirely separate industries. Canada is one of a select group of countries that can sustain itself on its fossil fuel resource base and contribute to the ability of the world to do so during a global transition to other forms of energy. Canada produced 3.15 and consumed 2.22 million barrels of oil per day in 2006 (BP p.l.c. 2007). There are many reasons to believe that the oil and gas industry in and beyond Canada can simultaneously address environmental concerns, continue to thrive, and show other industries how to accomplish this.

In the following section, we examine "The Opportunity for Global Leadership" available to the Canadian oil and gas industry—defining "global sustainability" and addressing "why the Canadian oil and gas industry" might be a particularly promising industry to look to for global leadership. Thereafter, we address "New Ways of Thinking and Acting That Can Make the Impossible Possible"—competencies and actions that companies everywhere can develop to help them to become environmentally sustainable:

- "An Inspiring Goal"
- "Thinking and Acting Capacity" embedded in three key competencies
  - o "Heedful Collective Thinking Focused on Sustainability"
  - o "Resilient, High Reliability Interaction and Response"
  - o "Intercompany Collaboration toward an Inspiring Goal"
- "Inspired and Inspiring Leadership"

Before concluding, we remind everyone of the criticality of "Getting Society on Board and Acting Positively."

Throughout this chapter we share evidence that environmental leadership can and does exist in the Canadian oil and gas industry by sharing examples of such leadership and relaying what we notice that gives us hope. Our approach is not value neutral. We focus on the positive. Positive, realistic, upbeat thinking underscores contribution, cooperation, creativity, and openness. People working toward a positive end are more likely to contribute above and beyond what they are required to do and to create a culture of cooperation. We are daily witnesses to success brought on by team members who believe they can find a solution and to failures brought on by those who do not engage prior to solution seeking.

The Opportunity for Global Leadership

The Canadian oil and gas industry has an exceptional opportunity to influence the world and the future. When it grapples with questions of intelligent discourse, reliable action, and collaboration and finds powerful and empowering answers to them, it will be moving itself, its international business partners, and society toward more sustainable ways of being and acting. It will also be offering at least a partial model of how the rest of the global oil and gas industry can do the same and how many other industries can follow suit.

Other organizations in other contexts will need to develop solutions that work within their situations. The fundamental principles will, however, remain: *Embed, Cooperate, Act*. (1) *Embed* thinking about sustainability into company mental and action models and into their decision-making systems. (2) *Cooperate* (and align) with other organizations (including competitors) to share and develop knowledge and to build industry-wide thinking and acting capacity. (3) *Act*: go beyond simply wishing and dreaming—and well beyond denying that such bold steps are possible—and simply engage in creating an environmentally sustainable industry.

The Possibility of Global Sustainability

Such phrases as "sustainability," "sustainable development," "environmental sustainability," and "global sustainability," are frequently used in addressing the integrated set of threat-to-life-on-this-planet problems that were suggested almost one-and-a-half centuries ago by George Perkins Marsh (1864) (as cited in Shabecoff 2001) and that have become increasingly apparent to many over the last half century (Meadows et al. 1972, 2004; United Nations 1987; McKibben 1989; The Nobel Foundation 2007).

This chapter pairs the word "global" with the word "sustainability" and uses two phrases to define "global sustainability." The first phrase modifies slightly the Brundtland Commission definition—"Sustainable development is development that meets the needs of the present without compromising the ability of future generations to meet their own needs" (United Nations 1987)—to define sustainable development as development that "meets this generation's needs in ways that *enhance* the capacity of future generations to meet theirs." The second phrase is "a world that works for everyone with no one left out" (Stoner 2006; Stoner and Werner 2006).

"Global" Sustainability

The use of the word "global" calls attention to two perspectives: the extensiveness of the "issue" of sustainability and the interdependent nature of the issue.

Extensiveness
First, the world has by now reached a state where achieving sustainability anywhere requires achieving it everywhere. The problems of pollution arising from the burning of coal in China are now causing health problems in California (Bradsher and Barboza 2006), and according to an article by Michael Casey (2007), "It takes five to 10 days for the pollution from

China's coal-fired plants to make its way to the United States, like a slow-moving storm." The acidic precipitation falling on Toronto is at least partially sourced in the American Midwest (Environment Canada 2007). The issue of sustainability is global in the sense that it involves all parts of, and everyone in, the world.

Nature

Second, the "issue" of sustainability is global in another sense—the sense of interconnected, integrated, complete, and whole. If the peoples of the world are to move to a way of being in the world that allows future generations to thrive, then all of the "pieces" of the world will have to exist in an integrated, harmonious whole rather than as the separate and independent components that some (seemingly) see them to be. Diseases that are endemic in one country may become pandemic around the world. These types of problems are not solved anywhere until they are solved everywhere. They are truly global.

The Brundtland Definition

Future generations

Perhaps the strongest aspect of the Brundtland definition is its explicit intergenerational message—the explicit call to consider the future and the peoples and species that will live in that future and to commit to creating a world that supports them.

Enhancing capacity

The substantive difference in our definition, relative to the Brundtland definition, is the word "enhancing." Adding that word is intended to raise the bar for the efforts of all of us by calling attention to the possibility of restoring and developing the environment and embedding this potential in our thinking and actions.

A World That Works for Everyone

The second definition we use, "a world that works for everyone with no one left out," may or may not add substantively to the first. It is intended to call attention to two things, the breadth of our view of what "global sustainability" can mean, and our emphasis on the "sustainability issue" as being not only about the environment (even though we focus on environmental sustainability in this chapter).

Works

The word "works" is intended to call attention to the total quality of people's lives, not just the quality of the air they breathe and the water they drink.

No one left out

The words "everyone" and "no one left out" are intended to call attention to and underline an appropriate interpretation of "this generation" and "future generations" in the Brundtland definition.

> A sustainable world is ultimately a just world that is environmentally, socially, politically and economically "healthy." It is a world within which people live a "good" life. They have food and water, and energy; are free of disease and can take this safety for granted. They thrive within communities that they are meaningfully involved in and that are themselves socially and economically sustainable. They are embedded within political systems that align with those of others to create a larger social order that is also sustainable. They thrive within organizations that are embedded in their communities.
>
> (Milton 2007)

Why The Canadian Oil And Gas Industry?

There are at least four good reasons why the leaders, managers, workers, and investors in the companies that make up the Canadian oil and gas industry are excellent candidates for demonstrating how companies in all industries can move toward a sustainable world. These are the same reasons that poise them to take a leading role (perhaps the leading role) or to work with others in shared-leadership roles to transform the oil and gas, energy, and other industries.

1. Canada Is a Legitimate Actor in the Oil and Gas Industry, Which Is a Global Industry

Sustainability initiatives developed by Canadian companies have the potential to be adopted globally. Canadian-based oil and gas companies are active in most major global markets, from at home in Canada and the United States to the Middle East, Asia, the North Sea, the Caribbean, and South America and again have an excellent reputation. Encana operates in France, Denmark (Greenland), Oman, and Qatar. Nexen operates in Yemen, Nigeria, Colombia, Indonesia, Brazil, and Equatorial Guinea. Canadian

Natural Resources Limited (CNRL) operates in the North Sea, Côte d'Ivoire, and Gabon. Other Canadian companies have operations that cover the globe operating in at least sixty countries.

Canadian industry is facing most of the major issues that will need to be resolved to create an environmentally sustainable oil and gas industry—Canada is developing oil sands, is active in deep-water exploration, and is involved in the Arctic. Water shortages, energy costs, and rising temperatures affect oil and gas producers like they do other industries. Growing populations and changing weather patterns force organizations and the industry to use new sources or increase recycle rates as their feed of fresh water is restricted. Energy costs force more efficient recovery methods in oil sands. Rising temperatures restrict the drilling activity in northern muskeg regions, which must take place on frozen ground.

2. The Industry is the Highly Visible and Environmentally Relevant Oil and Gas Industry

The oil and gas industry is highly visible in all its actions and all its impacts. In the oil and gas industry, what individual companies and governments do is the source of headlines, government inquiries, diplomatic actions, and international agency concern. The oil and gas industry is very transparent. Material events, including company operations, finds, spills, and other incidents—both positive and negative—are publicly released into the investment community almost immediately upon their occurring. What Canadian companies in this industry do will be noticed throughout the world.

Many Canadian companies have made substantial investments in sustainable business practices already. Some have voluntarily elected to achieve sustainability in every area of the company. Companies seek development of such practices not only as a means to differentiate themselves but as a better way forward. CNRL, in their international division, Canadian Natural Resources International, actively supports the International Chamber of Commerce Charter on Sustainable Development and, according to their website, are currently engaged in obtaining ISO 14001 certification. Nexen issues an annual Corporate Sustainability Report in which it reports not only on successes but also on disappointments the organization experiences.

In this industry, companies have a particular need to have their actions recognized and endorsed by governments, regulators, and the communities in which they operate. By issuing their own report cards, these companies present themselves for scrutiny and endorsement which, when obtained, provides encouragement for others to subscribe and opens the nonsubscribers to public criticism. By commenting openly and proudly about successes and disappointments, companies like Nexen also demonstrate a commitment to learning from their own and others' experiences.

Sustainability initiatives will become diffused globally as Canadian companies import them to the countries in which they operate or require their global partners to use/commit to environmental sustainability.

3. Canada Is a Boundary Spanner

Boundary spanners unite, reconcile, or transfer knowledge between groups (Thomas-Hunt and Gruenfeld 1998). As a legitimate actor in the global oil and gas industry and as an experienced, highly regarded country, Canada and its industries are well positioned to help diverse constituents in industry and governments to communicate and work together effectively. By modeling what is required, it may inspire others to play similar roles.

Canada's unique history, diplomatic reputation, and multicultural population will help it and its industries in these bridging roles. Canada is generally trusted and held in high regard. It may, in fact, be in a relatively unique position of being able to unite North America with countries across both the Atlantic and Pacific—simultaneously strengthening TransAtlantic relationships by brokering between the United States and Europe and relations between other countries and Asia (Blair 2007). In 1970, Canada (which already had strong relations with Europe and the United States) was the first industrialized country to open diplomatic relations with China (Foreign Affairs and International Trade Canada 2007). Other countries followed.

4. Why Not?

The most important reason for the members of the Canadian oil and gas industry to provide leadership in protecting the industry's own viability and in providing a model for all industries to learn from, modify for their own circumstances, and move forward with is that there is no valid reason not to do so. Legislation is creating an environment in which companies need to develop or adopt business models that create environmentally sustainable, economically viable companies. Those that do so first—the leaders—create business opportunities.

Companies and company consortia may particularly benefit by embedding their commitment to environmental sustainability within themselves and by delivering clean energy that exceeds current legislative requirements. Those at the top of the industry learning curve should be less susceptible to continuously having to scramble to meet ever-tightening legislative requirements. They may even garner influence that helps them to become leaders whose approaches to environmental issues are institutionalized within the energy industry. Gaining a solid reputation for being environmentally sustainable and having astute puzzle-solving skills may help them to attract highly technical personnel who then contribute and help the companies to continue to lead.

New Ways of Thinking and Acting That Make the Impossible Possible: An Environmentally Sustainable Oil and Gas Industry

The oil and gas industry is a global, cost-based, commodity industry. By and large, oil is oil and gas is gas. Companies compete on access to resources, on reliability of supply, and on margins. They develop and maintain competitive advantage using firm-specific resources and capabilities (Wernerfelt 1984) that help them find, secure, extract, process, and distribute oil and gas as inexpensively and as reliably as possible.

The industry is highly technical and dependent on knowledgeable workers, research and development, and execution. New technology that lowers costs along the value chain or reduces uncertainty in exploration, production, or distribution is highly valued. Technology often diffuses through the industry via licensing agreements or via imitation. Early adopters have a temporary advantage until others adopt, imitate, or adapt.

The global environment for oil and gas is both changing and turbulent. Geopolitics is complex and constantly morphing (Tertzakian 2007). Public and industry concerns about environmental stability complicate this already complex industry. This is a challenging context within which firms need to think and act carefully in all areas of their business; it is a context within which companies, the industry, and society can benefit from cooperating with one another. *We focus on thinking and acting capacity and on across-company collaboration.*

At a basic level, to be successful, companies need to understand their industry and the nature of competition therein. This knowledge-based foundation is, however, insufficient. Research suggests that the companies that excel will be those that think, act, and learn in real time (Brown and Eisenhardt 1998, Huber 2004). We argue that industry leaders in sustainability will be the companies that (1) figure out what they need to do and actually do it as knowledge and circumstances evolve technically, procedurally, and politically across interdependent constituents; (2) reliably implement what they have learned, while still learning—from their own and others' successes and failures; (3) take intelligent risks; and (4) cooperate with one another to learn about and embed sustainability in their companies and across the oil and gas industry. Companies that and people who truly excel will be those with the capacity to see the entire system unfold as it happens; these will be those with the capacity to see the entire game board, understand interdependencies therein and plan and act while anticipating how their actions will affect others, and developing contingencies that take this into consideration. This level of thinking will provide a solid base for them to continue to act calmly and intelligently when others act in unexpected ways or when unanticipated events occur.

This straightforward description is deceptively simple. It is easy to understand and often difficult to achieve, particularly when companies feel threatened as their industry environment changes in unexpected, unpredictable, or misunderstood ways. In these circumstances, instead of embracing change, companies (and industries) often become stuck in the status quo digging in and continuing to exist in familiar, rather than effective, ways.

Nevertheless, there are reasons to be optimistic. The oil and gas industry has a history of successfully morphing as circumstances change (Tertzakian 2007). The thinking and acting capacity within the industry is notable. The highly technical and knowledge-based labor force of many companies is used to operating in difficult circumstances, resolving unforeseen issues as they arise, and anticipating and preventing others from doing so. Mistakes can be costly in terms of money, opportunity, and reputation. Many, as we have seen, have already shown an interest in and commitment to sustainability.

Research suggests that companies that envision themselves as sustainable and that develop their thinking capacity to figure out how to do so in real time will be the ones that succeed.

An Inspiring Goal

Ways of thinking that can move any of us toward making the impossible possible often start with an inspiring goal, which unleashes capacities to think and act in new, unfamiliar ways. Inspiring goals call forth inspired leadership and widespread support and action.

Instrumental, economic arguments are widely used to convince companies to take sustainability seriously. Companies are being "educated" about how to improve their performance and gain a competitive advantage by identifying and acting upon corporate social opportunities. These arguments are often legitimate and appealing to corporations and their shareholders. They often motivate companies to introduce sustainability initiatives (e.g., Grayson and Hodges 2004; Willard 2002). They can, however, also be dangerous. Even though companies and shareholders may benefit and the environment may be improved, the economic rationale and self-interest that motivates action may ultimately provide a self-defeating logic. While the language may be familiar, companies are being asked to insert a counterintuitive frame into their own economic model. Not all will be convinced—they may see such arguments as the latest sales proposition. Those who are convinced may remain committed only as long as the business case for sustainability holds (see Ferraro, Pfeffer, and Sutton 2005, for a related discussion of how economic thinking affects corporations). Competing logic about why and how to proceed may introduce confusion that results in well-intended incremental or misguided actions that undermine the effectiveness of sustainability initiatives. Initiatives may be poorly thought out, or the point of a change may be missed in narrow, myopic thinking. *Most importantly, this approach misses the opportunity to frame and act on sustainability simply because it is the right thing to do—and thereby it misses the opportunity to rally human energy around a truly inspiring goal and to achieve the excellence that accompanies this frame.*

The goal of creating an environmentally sustainable world that meets this generation's needs in ways that enhance the capacity of future generations to meet theirs: a world that works for everyone with no one left out is a goal worth getting up in the morning to contribute to. It is specific enough to provide direction yet broad enough to encourage imagination, instill hope, and inspire people to work together to achieve it. It is positive, achievable, and meaningful (Locke and Latham 1990). Its positive orientation and relevance make it attractive. It is challenging yet realistic. There are reasons to believe the oil and gas industry can succeed in attaining this goal. It is a goal that can provide a rallying point for individuals and organizations—a goal that can strengthen the thinking and acting capacity that companies need to become and remain environmentally sustainable.

"New Thinking and Acting Capacity"

To make wise choices and become environmentally sustainable, companies will need to process considerable information in potentially unfamiliar ways. They will need to decide when and how to act and when not to act in what may seem like a split second to those who are used to operating in more stable environments.

On the basis of information that may seem insufficient, they will need to make intelligent new commitments, abandon current courses of action, or do nothing even when others are in action mode or sometimes when key stakeholders are demanding action. The capacity of a company to think carefully and act wisely in a changing and often threatening context cannot be taken for granted. For good reasons, the most astute oil and gas companies may have difficulty determining what (if anything) to do to become environmentally sustainable.

As noted, the companies that excel will be those that have developed their thinking capacity. They will be the companies that intelligently scan and probe their environment for major threats and opportunities in a reliable and timely way. They will be those that process information quickly and inclusively so as to develop their collective ability to pay attention to what is happening, make decisions, and follow through—again in a timely way. In the language of research, these are companies that think and act heedfully. They have well-developed heedful, collective minds and action patterns. They think and act resiliently in highly-reliable, intelligent ways. All are necessary and, we argue, are strengthened when companies cooperate with other organizations (including competitors) and thereby embed sustainability across their industry.

Heedful Collective Thinking and Action That Is Focused on Sustainability

Companies that think and act heedfully have well-developed, heedful collective minds and action patterns. A heedful collective mind is a multiperson mind that pays attention to what is happening and makes wise decisions based on current information. When a company is thinking and acting heedfully, each action is dynamic in its execution and is thus modified by what occurs just before it. When, in contrast, an organization thinks habitually, it thinks in terms of what it would have paid attention to in the past. Each action then becomes a replica of its predecessor (what the organization would have done in the past). Habitual thinking and acting is problematic in an industry context where what is important changes. The same actions occur across time even if the context has changed. Refer to Weick and Roberts (1993) for a related discussion.

Heedful thinking can be a challenge for organizations that are used to paying attention to certain kinds of information and not others. The information that they are used to considering may become more or less important when circumstances change, or it may become more or less important to people who do not normally receive it. New types and sources of information that companies do not routinely, or ever, monitor—let alone consider—may become critical. The company may not realize any of this. Systems and structures that support paying attention to what has always been paid attention to and providing this information to people who have always received it may not adjust.

To the extent that the information that companies need to pay attention to (e.g., sustainability, the oil and gas industry, technology, company commitments) is distributed among people throughout the organization, heedful thinking must occur across the organization—collectively. When a collective, interdependent sense-making system is functioning effectively, the people involved share relevant information in real time (across disciplines and languages). They know what they are talking about (are up-to-date in their areas of expertise), know what they do not know, and have excellent judgment in knowing what is important and when and in sharing this knowledge when it is important to do so. They know who needs to know what and they understand how to work a thought through a project or organization. They have impeccable timing, knowing when and where to speak out, how to be heard, and how to follow-up effectively. Their thinking skills and understanding of broad objectives are coupled with a high degree of self-awareness.

Organization members in these systems are able to rely on one another for different types of information. The group as a whole cooperates to think through and integrate disparate points of view (sometimes using a range of communication media because individuals are geographically dispersed). Debate, candor, and listening characterize discussion. People focus on ideas rather than on attacking one another. Changes in knowledge and context are considered in a timely way. As ideas are implemented, the group continues to learn and to modify what the company is doing when it makes sense to do so. Those involved recognize long design and implementation cycles and make intelligent decisions about when costs are sunk and when change in current projects is imperative. New ideas and lessons learned are implemented across time horizons and projects. Lessons learned in one context are transferred to others. Learning is important in complex technical fields where moving forward is seldom about two choices—A or B, Black or White, Right or Wrong—fields where all choices become a matter of trade-offs.

When errors occur, organizational members learn and use their new knowledge to adapt to changes in the organization's environment (Sitkin 1992). They use failure as a basis for examining what they infer and think and what this suggests for solutions. Over time they will become more adept at and comfortable with recognizing, interpreting, and responding to events when things go askew—as they often do when people and organizations are pushing the limits of what they know. Over time the organization develops its knowledge base, which itself often becomes a source of competitive advantage. Individuals within the organization develop a deep understanding that helps them to improvise and act independently in a sensible way as circumstances change.

An open system of information sharing and development is held together by trust. This means that individuals within the system need to be trustworthy themselves (competent, reliable, emotionally stable, ethical) (Milton and

Milton 2000) and need to be capable of trusting others. From this foundation, individuals share information and indicate the limits of their knowledge. They do so openly and hold themselves accountable for their comments and actions. The individuals involved are emotionally stable or centered enough to perform well and to help the group perform (Druskat and Wolff 2001). While idiosyncratic, sometimes odd behavior may occur, the general atmosphere is such that people can take risks without fear of reprisal (Golembieski. 1989).

Current hierarchical systems and people with outdated mental models of power who protect their information because they believe that information is power are often barriers to intelligent and heedful thinking across an organization. In industries that depend on collaboration where organizations need to develop their collective sense-making abilities, the ability to share information becomes a source of power. By sharing information, individuals become central in sensemaking networks. Different people lead in different areas and individuals seize opportunities to contribute to and even lead in the areas that they know about and are interested in. (Refer to Tapscott and Williams [2006] for a related discussion of mass collaboration.) Higher-level leaders in collaborative organizations focus on leading the system, for instance, by ensuring that people have the resources (including information) that they need, that the right people are working together, that the thinking and acting system itself is functioning effectively, and that it is connected to its environment but buffered from deflections. As heedful interaction increases, organizational errors decrease (Weick and Roberts 1993).

The ability of companies to maintain a heedful mindset as they process and act upon information as circumstances change must be nurtured. Most companies will need to learn how to develop and sustain this strategically relevant capability. The issues involved are complex (involve multiple disciplines and ways of thinking, are connected to people's value systems of what business should and should not do, and may require letting go of approaches that appear to have been successful in the past). The capacity of humans to process information flexibly as events morph is limited. Complex, ambiguous, and changing circumstances challenge dispassionate, reflective thinking and action. Challenges become more apparent when individuals need to process information quickly in contexts that are fast-paced or that depend on the ability of people to work interdependently.

Think of a person as having a limited number of cognitive resources and of these resources as being affected by how much there is to pay attention to and by physical limitations. Think also of how individuals react to turbulence and fast-paced environments and how this affects their relationships with one another. As fatigue and stress set in, they may argue more or become less civil to one another or they may withdraw and not go that extra mile for the group. In an effort to cope, they may become more entrenched in their roles at the exact moment when organizations need them to think across knowledge bases to develop points of view and action plans that capture the strengths and avoid the weaknesses of any one person.

In changing, ambiguous, complex situations, people tend to become more rigid in their thinking and action. (Staw, Sandelands, and Dutton 1981). They tend to see alternatives narrowly emphasizing prior expectations and internal hypotheses, focus on dominant and exclude peripheral cues, and respond in well-learned and habitual ways. In the process, they may rely too much on their previous training, escalate existing commitments, and become more risk averse than usual, often preferring the status quo to change. In these and other ways, they may try to create certainty beyond that which exists and in so doing undermine the quality of company thinking and action—increasing the likelihood that they will respond inappropriately to their situation. Being vigilant and comprehensive may make them feel better in their changing—perhaps uncomfortable—context. Illusions of stability and predictably may mislead them and undermine company performance. Comprehensive decision making is negatively associated with firm performance in an unstable industry (Frederickson and Mitchell 1984). Sunk costs and past commitments to courses of action that no longer make sense may both motivate and flow from faulty thinking.

A broad, visionary corporate goal focused on sustainability will help develop heedful, collective thinking and action focused on sustainability within the company. Knowing that sustainability is important, company members will be more likely to think and speak out about sustainability. Critics will be less able to defeat good ideas and disenfranchise their proponents. A broad goal and general guidelines will encourage expansive thinking and action. Broad, rather than hierarchical, collaboration will be more likely to follow.

Resilient Thinking and Action

Resilience (the capacity to rebound from adversity strengthened and more resourceful) is at the heart of positive organizing (Sutcliffe and Vogus 2003) and, therefore, is also likely to be at the heart of long-term change toward an inspiring goal—a goal such as global sustainability.

Although there is considerable evidence to show that organizations often narrow the way they process information, tighten control, and conserve resources when they feel threatened (e.g., in uncertain contexts), they do not always do so. Faced with adversity, organizations (and individuals) may respond resiliently rather than rigidly (Sutcliffe and Vogus 2003).

Resilient thinking and action results in broader information processing, loosening of control, and the effective use of slack capabilities and resources (Sutcliffe and Vogus 2003)—exactly the behaviors that would support the heedful thinking discussed previously (and intercompany cooperation—a behavior considered in the next section). By creating a context that supports resilient thinking and action, companies should be better able to meet the challenges that environmental issues present.

Individuals and organizations have a greater capacity to respond to circumstances they perceive as threatening when they believe they are doing well and when they have previously achieved good outcomes (Masten and Reed 2002). Both will be more likely when the organization thinks heedfully and has a history of sound judgment. Organization members who have faith in the organization's ability to think and act capably in tough situations will be more likely to help the organization to do so. Their action will be fueled by their sense of group efficacy. They will be like the members of a technology company who, when asked by one of the authors whether they knew anything about mergers and acquisitions, said with confidence, "No, absolutely nothing, but we will figure it out. We do this figuring out well," and with that they proceeded to grow their company very successfully through mergers and acquisitions.

As individuals and areas within an organization take action to advance sustainability, they will become more committed to doing so. Heedful thinking and action focused on sustainability will become more resilient and reliable—embedded—as it becomes normalized within the company.

Many oil and gas companies will already have confidence in their ability to solve challenging puzzles. Their capacity to effectively address technological and other challenges suggests that they have the potential to execute reliably in tough situations. Their capability to alter their mental models, change, and make effective choices as circumstances morph in ways that fall outside of the parameters they are used to considering may be more of an issue. As noted earlier, many are also already committed to the environment and have been successfully making a positive difference. Many others are trying. Publicizing the good news stories would help organizations to feel good about what they and the industry are doing, to feel optimistic about their success, and to feel that they have some control over their destiny in advancing sustainability issues. As a consequence, they may be more open to embracing a commitment to sustainability and the changes its pursuit will require. Self-esteem, optimism, and perceived control are associated with individuals being open to large-scale organizational change (Wanberg and Banas 2000). People are more likely to become attached to and stay with their organization and contribute to sustainability and the changes required for it when the organization is respected and accepted in the larger community (Dutton, Dukerich, and Harquail 1994).

Many recognize that resilience and highly reliable actions are needed in emergency-response groups and in space-shuttle crews. What fewer realize is that these same capabilities are required elsewhere, including in oil and gas companies.

The thinking capacity of an organization is a valuable resource. The idea is to embed thinking about sustainability into the mental model of the organization, so that sustainability issues are routinely, even automatically, considered in an intelligent way.

In the next section, we address the third capability that we argue oil and gas companies should develop to help them become environmentally sustainable: their capacity to cooperate with other organizations (including competitors).

Intercompany Collaboration Toward an Inspiring Goal

The issues that need to be addressed to create environmentally sustainable oil and gas companies and a sustainable energy industry are far too encompassing for any single organization to resolve unilaterally. Many minds are needed to solve the technical, business, and political puzzles that will lead to environmentally sustainable fossil fuel extraction and processing and the development of nonfossil fuel-based, renewable sources of energy. In this section, we encourage companies to cooperate with one another and with other organizations to embed sustainability both within companies and within the industry.

Cooperation and stronger forms of collaborative alliances are often recommended as one way for firms to cope with turbulent and complex environments (Gray and Wood 1991). Organizations cooperate when they act in a coordinated way to pursue shared goals, enjoy a joint activity, or further their relationship (Argyle 1991). In so doing, they may pool their assets and capabilities and form collaborative alliances on multiple dimensions.

Pooling resources and responding to threats are two common reasons for establishing cooperative efforts (Axelrod 1984). Both suggest that oil and gas companies should cooperate to help one another, and the industry as a whole, become environmentally sustainable. Linkages to other organizations may help firms (Oliver 1990) and the industry to survive.

Cooperating will help firms and the industry to think intelligently about sustainability and to decide what to do. Their collective sensemaking will serve as a foundation for their working together to develop technology and for their working productively with society (e.g., governments, not-for-profit organizations, communities) to address existing environmental concerns and, over time, to develop creative ways of enhancing the environment over and above merely stopping its degradation. By pooling expertise, industry will be more likely to succeed.

There is a structural need for companies to cooperate. One or two simply can't do everything. Given the ambiguity and complexity of the issues, it is unlikely that they could do so even if they wanted to. Expertise is distributed; time is short. The magnitude and complexity of issues that must be addressed can be daunting. Interdependencies among actors require implementing solutions across people, organizations, nation states, and markets. These are circumstances that require and support mass collaboration. Refer to Tapscott and Williams (2006) for a related discussion.

Canadian-based companies have experience working across the globe and in various sectors of the industry (e.g., conventional oil and gas, deep-water drilling, heavy oil). Their unique and collective knowledge will be valuable, arguably essential, to help the global oil and gas industry become environmentally sustainable. None of them is likely to have the expertise, experience, and other resources to develop industry-wide solutions for environmental sustainability under existing time constraints. Companies and individuals with diverse and required knowledge and experience will need to work together. Highly complex tasks generally require that considerable information be processed (Kelly, Futoran, and McGrath 1990) using a range of knowledge and skill bases (Wood 1986). When circumstances are changing and outcomes are ambiguous, considerable judgment is often required; existing protocols may not apply.

By cooperating, companies could collectively underwrite the costs of developing sustainability solutions that they would otherwise incur alone. Although individual companies may be genuinely motivated to develop solutions on their own, they also need to survive economically in their current cost-based commodity business. Investing in straightforward, immediately profitable sustainability opportunities is a simple choice; everyone wins quickly. Yet investing too much or too little may have negative effects. On the one hand, investing too much in sustainability-oriented technologies that involve long and uncertain payback periods may cause management to lose support in the financial markets that sustain them. On the other hand, not demonstrating a commitment to the environment may damage their reputation with the public and other internal stakeholders (as experienced by the New York Port Authority before it became more sensitive to the needs of the homeless [Dutton, Dukerich, and Harquail 1994]). For many executives and their companies, the sustainability issue may be becoming a question of "how we can do what we know is the right thing to do, when the financial markets punish any action that does not clearly and unambiguously contribute to raising today's share price."

The financial markets may be more "supportive" if the entire industry publicly and collaboratively commits to environmental sustainability and if the public and their institutions respond positively. In the aftermath of crises like the Exxon Valdez, Alaska; Union Carbide, Bhopal, India; and Chernobyl, Ukraine, disasters many have already begun to factor environmental risk assessments into their investment decisions. By working together, companies can share the costs, lobby for public moral and financial support, and gain momentum in greening their industry.

To increase the likelihood that interfirm alliances will succeed, each of the participating firms should have solid, well-communicated, and understood reasons for doing so. Firms with different objectives may commit to a course of action (Lindblom 1959) and work together effectively as long as the total set of objectives across firms is compatible and transparent. Effective cooperative alliances are those within which partners value cooperation and interact cooperatively. Alliances that include mechanisms and structures that support cooperation are more likely to succeed. The focus in cooperative alliances is on partners working together toward a common goal and pooling assets and capabilities in the process.

As already discussed, an inspiring, overarching goal itself would rally support and cooperation. Achieving milestones would similarly do so. The process of pooling resources would imply motives of reciprocity that are likely to emphasize cooperation (Oliver 1990). Achieving milestones would increase optimism and a sense of efficacy in the alliance. Leaders would themselves serve as exemplars of cooperation (e.g., by being generous and adopting a long-term view of relationships). Their status and record would inspire others to join.

Each of these factors contributed to the success of SEMATECH, a consortium of U.S. semiconductor manufacturers and the U.S. government that was created to pool resources and reestablish the U.S. semiconductor industry (Browning, Beyer, and Shetler 1995). SEMATECH was jointly founded by fourteen firms that together accounted for 80 percent of the semiconductor industry. Although all firms in the industry were approached, not all agreed to join. Participating firms contributed $100 million annually, funds that were then matched by the U.S. government. Member companies were involved in consortium decisions at SEMATECH from the beginning. Although, insufficient attention to implementation and the process of managing alliances has prevented many from achieving their potential (Gulati and Nohria 1992), this problem did not dominate SEMATECH. Governance structures, such as a board of directors, comprised of high-level directors from member companies and technical boards who approved and advised on projects. Organization culture was aligned with cooperative achievement. Individuals were expected to cooperate with one another.

Although firms may work together when they have no other options, cooperative behaviors tend to emerge when partners feel safe in an alliance. They trust one another, and behaviors and structures reinforce their trust. Each believes its goals will be met and that partners will behave cooperatively rather than opportunistically. Those involved deliver on what they promise and a norm of reciprocity emerges. Trust thus tends to emerge over time and to depend on the quality of the relationship between the firms involved. Firm practices that establish trust can have a major impact on cooperation and the eventual success of an alliance. High trust enhances alliance effectiveness and facilitates ongoing relationships between firms (Dodgson 1993; Sharfman, Gray and Yan 1991).

A belief in cooperation is more likely to emerge when partners view alliance relationships as continuing and perceive all concerned to benefit (Axelrod 1984; Beyer and Milton 1994). On the one hand, given the entrenchment of a competitive ethos in the oil and gas industry, a complete paradigm shift to cooperative thinking may be viewed as unlikely. On the other, there is a history of cooperation in the industry that suggests firms will work together and do so effectively.

Core values that favor individualism, achievement, and competition emphasize an overriding value of self-reliance, which may be seen as almost antithetical to cooperation. While there may be a tendency to focus on penalties for misbehavior, focusing excessively on creating disincentives for cheating during the initial contracting phase frames alliances negatively from the beginning and may breed distrust and block cooperation. Negotiating every detail of a contract does not guarantee success (Bleeke and Ernst 1991) and may undermine it by imprinting distrust and tit-for-tat thinking in the minds of those involved. Oil and gas company alliances that are implemented and managed in a way that normalizes cooperation will be much more likely to succeed. Broad agreements governed by general principles and information sharing together would encourage the genuine collaboration needed (as long as companies felt safe and protected because they trust the other parties). Alliances that value cooperation and that interact cooperatively will be characterized by structures that support cooperative interfirm behavior. Cooperation among competitors will require courage and an inspired goal that people commit to. In some respects, the mindset is akin to a redefinition of the commons where ownership of knowledge and acting capacity is altered.

Companies may cooperate with one another directly or via industry bodies to increase their capacity to think and act heedfully and resiliently. In Alberta, industry associations such as CAPP, PTAC, CONRAD, and others actively bring together companies, universities and government agencies in efforts to collaboratively solve the challenges. The Canadian Association of Petroleum Producers, CAPP, publishes an annual Stewardship Progress Report that highlights industry successes and provides hard statistics for 103 member companies comprising 99.98 percent of CAPP member production, which itself comprises 97 percent of Canadian oil and gas production. This report contains numerical summaries of safety records, greenhouse gas (GHG) and other emissions, spills, and water use and is publicly released. CAPP promotes a stewardship program and assists new industry players in the area of stewardship.

The Petroleum Technology Alliance of Canada, PTAC, facilitates collaborative research and technology development in a number of areas and reports that of the $133 million spent in joint industry research through PTAC, a full $98 million has been spent on projects in the areas of the environment, efficiency, and sustainability. In 2006, they launched their Technology for Emission Reduction and Eco-Efficiency (TEREE) Phase II project with a focus on emission reductions and efficiency improvements. Water use came under the increased focus of PTAC in 2007 when they hired a dedicated Director of Water Use to focus on the need to reduce industry water use and use water with greater efficiency. Suncor reported in their own sustainability report in 2007 that they see water as a major constraint (Suncor 2007).

Suncor and other major oil-sands producers, including Syncrude and Shell, are members of CONRAD, the Canadian Oil Sands Network for Research and Development. This group is highly focused on the environmentally sensitive issues surrounding water use and fine tailings. Each year members and interested parties meet to discuss and promote ways for reducing water usage in their processes.

Industry initiatives such as these are important to developing the thinking and acting capacity embedded in heedful collective thinking focused on sustainability and resilient, high-reliability interaction and response. Intercompany collaboration toward an inspiring goal is essential to each. In the next section, we examine the contribution that leaders can make to develop collective thinking and acting capacity.

Inspired and Inspiring Leadership

Leadership plays a pivotal role in instilling and developing a heedful collective mind and reliable action within companies. Intercompany collaborations devoid of leader support are often doomed to failure.

Through their actions and words, leaders exemplify what is important. Individuals follow through and act in ways that leaders exemplify and encourage. If leaders want members of their organizations to move boldly into ambiguous circumstances and figure out what to do on the basis of a general goal—such as becoming an environmentally sustainable company—these *leaders must step forward and commit their companies boldly and unequivocally to becoming environmentally sustainable and to doing whatever it takes to create an environmentally sustainable oil and gas industry.*

At that time and thereafter, the *leaders will need to encourage heedful thinking and reliable action at the same time that they encourage risk taking, experimenting, and learning from mistakes.* One way of doing so is by rewarding good tries, incremental improvements, and learning in the process. If all scientists who were trying to discover a cure for cancer were publicly whipped each time they failed to find a perfect solution, many might become less enthusiastic and less engaged in their work. Some might even change careers or discourage promising recruits from joining the "team" or both. Yet, if the scientists continued and found a drug that cured cancer and if that cancer later morphed, we would want these very people to be the ones who collectively examined the boundary conditions of their prior cure and pooled their efforts to find a new cure in the new context.

Inspired leaders have a big objective—an audacious goal that inspires people. They commit publicly to this goal and to doing what is necessary to achieve it, and beyond supporting the goal, they follow through. At the organization system level, they provide resources and create structures and mechanisms to provide ways for people within the company to achieve the goal. Particularly successful leaders may embed sustainability in their company culture by talking about it, sharing related stories, celebrating it at company meetings, normalizing candid talk toward a sustainable future, and creating ways for people to work together to advance it in the company and across the industry. Leaders have opportunities to imbue their organizations with meaning and related action (Trice and Beyer 1988). Leaders can make a very positive difference by embedding environmental sustainability in the company culture (via ideology, symbols, organization practices, stories and narratives, norms and socialization).

The unequivocal support of CEOs (and boards) for the vision of creating sustainable energy companies will be required to inspire across-company cooperation and collaboration.

Leaders who decide to encourage members of their companies to walk on the bridge deck while the bridge is being built will need to create the conditions that help those on the bridge to trust that they will get to the other side. There is no easy answer to what it will take for oil and gas companies to become environmentally sustainable; the knowledge base required to figure it out is distributed across multiple people and organizations. In this context, to be effective, *leaders will need to simultaneously lead from the top (as many may already be adept at doing) and create the systems and structures that distribute leadership across their companies.* There is no rule book to tell us how to act in the future. Leaders need to lead on the basis of general principles and provide ways for others in the organizations to decide how to act in circumstances that cannot be anticipated and planned for in advance. Companies need to invest in the minds rather than in the obedience of people throughout the organization. Even the military, arguably the icon of a command-and-control style of leadership, is instilling distributed leadership in its troops (Dallaire 2004).

Getting Society on Board and Acting Positively

Although many communities are extolling the virtues of sustainability and also taking action to become sustainable, society as a whole needs to do more and do what it does differently. Society and industry need to cooperate in order to make environmental sustainability a reality.

The Present

Need aside, cooperation is currently challenged by the structure of debate about environmental sustainability. We perceive industry and society to be focusing largely on the negative, drawing lines in the sand, and concentrating on threat, fault, and loss. This mind-set is understandable; it is also predictable. Change is hard. The stakes are high. No one wants to risk being put out of business. No one wants to be seen as being responsible for current problems. However, is there anyone who wants to be seen as self-serving to the detriment of the planet?

In many respects, society is (at the moment) simultaneously placing environmental concerns on the public agenda and calling for industry reform while supporting an economic system that is insensitive to the environment. Society and industry are thus locked into a physical and mental structure supported by the use of physical resources that are eroding the ability of the planet to sustain carbon-based life—including human beings.

On many sides, threats—real and imagined—appear to be contributing to rigid, relatively narrow thinking and action. In the age of mass media, blogs, and the like, hope and fear, especially fear are offered as "opinion." This opinion is then acted upon by politicians and others until real thought and debate is eroded or is no longer present. These actions increase uncertainty in an already uncertain energy industry. Industry executives may intelligently be living in fear of irrational rule changes. Subgroups representing different perspectives are forming, arguing with, and demonizing one another. All this is happening at the very time that open, intelligent debate, sense making, decision making, and cooperative action are needed.

Companies and societies often struggle to represent their own agendas in a situation where the collective situation requires attention. The same urgency and vigilance that is increasing awareness of environmental sustainability may polarize opinions and entrench issue-based coalitions. As is often the case in threatening, complex, relatively ambiguous situations, those involved often demonize one another—pointing fingers, assigning blame, and demanding action from one another. In doing so, they waste valuable time and erode their own energy and their ability to work productively together.

Industry and society are one and the same—a house too often divided. We are past the point of empty rhetoric and pointing fingers and discussing what others should do.

Now

Just as industry needs to develop heedful, resilient thinking and action focused on sustainability, so must society. The two must think and act in tandem to achieve the bodacious, visionary goal of environmental sustainability on this planet.

The Future

Society can do much to help industry learn how to become environmentally sustainable. Society could, and we would argue should, make a particular effort to help and to encourage and make it possible for companies to think and act heedfully and resiliently toward environmental sustainability. By working with industry, by investing in the thinking and acting capacity of industry, and by developing its own thinking and acting capacity, society can create a context that encourages companies to invest their brain power and acting capacity in sustainability. Removing barriers and providing incentives would be a brilliant place to start. Focus on potential. Celebrate success. Create a positive and structurally sound path forward. In each of these ways stimulate the contribution, cooperation, creativity, and openness that (as noted earlier) a positive, intelligent frame inspires.

Rather than demonize those who fail, society could, for instance,

- recognize those who are trying, tolerate or even celebrate inevitable failures, and help those who fail to learn from their experiences;
- stimulate public debate;
- listen with an open mind to what industry members suggest and invite them to recommend what society can do to help;

- publicly recognize (as previously discussed) that many domestic companies are already making serious efforts to be environmentally sensitive and many companies with headquarters outside of Canada have good records for environmental sustainability: Shell, British Petroleum, Statoil, BG, and Chevron are among these (see, e.g., BP p.l.c. 2006; Chevron Corporation 2007; Royal Dutch Shell plc. 2007; Statoil ASA. 2006; Urdal 2007);
- create more opportunities for industry and for industry and society to collaborate to share ideas and develop and implement solutions;
- stimulate intelligent debate about what can be done;
- underwrite more of the costs involved and encourage sharing solutions across industry (including across competitors);
- provide easier access to research and development funds and tax credits to stimulate the discovery of what it means to be sustainable and how that can be achieved;
- encourage exploratory, probabilistic research that creatively explores sustainability and develops technology;
- fund intelligent best tries that may seem bold and based on leaps of faith by current standards while still continuing with what may seem like safer projects; and
- integrate social and physical science criteria and tolerances for error to encourage good tries.

A research and development system, based on criteria that are anchored to what has been successful, historically runs the risk of replicating the past, moving too incrementally, or building knowledge too slowly, rather than creating a bold new future. Just as comprehensive thinking undermines industry profitability in changing contexts, so may it undermine the potential of science to be intelligently flexible and make judgment calls as events unfold.

In these and other ways, we suggest that society invest more in the infrastructure that is needed to stimulate collective sensemaking and puzzle solving to achieve sustainability. As it helps industry to develop its thinking and acting capacity, society will strengthen its own capacity in these areas. In so doing, society will face the same set of challenges that research suggests industry will face. As previously discussed, the thinking capacity of human beings is often undermined by change, complexity, and ambiguity—especially under conditions of threat.

Canada, a country with one of the most educated populations in the industrialized world, has the opportunity to be a leader in knowledge development and in creating domestic and international systems to advance this. It could lead the way in developing and transferring new technology and education to the industry by dedicating brainpower to this task and by developing structures and systems that encourage industry collaboration.

Although we did not emphasize the business logic, there is a business case for sustainability. Even though we have argued that changes made for business reasons may too easily be abandoned when the business case changes and that economic logic is thus fickle, we clearly must recognize the business context of the oil and gas industry.

Recent research has found that many companies with strong financial performance invest in their societies and that doing so is positively associated with their subsequent financial performance (Margolis and Walsh 2003; Orlitzky, Schmidt, and Rynes 2003). These relationships may be a consequence of society investing in socially responsible companies that give back or of effective leaders creating more profitable companies that invest in societies. Or, they may be a consequence of both a committed public and committed leaders.

The authors conclude, and we reiterate, that the economic shareholders of firms often benefit from the contributions firms make to society. So, although in this chapter we have not focused on the business case for sustainability (and in fact have argued that this ought not to be the primary rationale for change), there is a business case. Even so, it is not clear whether investors in current financial markets invest in companies that are environmentally sustainable to the extent needed. Those who are committed to creating an environmentally sustainable society that survives them can demonstrate their support by investing in environmentally sustainable companies and by developing the thinking and acting capacity of industry and society with respect to sustainability.

Conclusion

This chapter focused on creating sustainable companies. It began by asking and examining three questions that would require a willingness to go beyond our current levels of thinking. The questions were: (1) How can a complex industry create the conditions that will encourage its multiple constituents to work together to create a sustainable world? (2) What can these constituencies do within and between themselves to create and sustain intelligent discourse about alternatives and to make and implement intelligent far-reaching decisions? (3) How can we get started doing what needs to be done?

The fundamental premise of our argument is that we—industry and society—have to *Embed, Cooperate, and Act*: (1) *Embed* thinking about sustainability into corporate and societal mental and action models and into decision-making systems. (2) *Cooperate* (*and align*) with other organizations (including competitors) to share and develop knowledge and to build system-wide thinking and acting capacity. (3) *Act*: go beyond simply wishing and dreaming—and well beyond denying that such bold steps are possible—and simply create a sustainable world.

Sounds easy? Think again. Science suggests that we will have to be deliberate and intelligent to develop our capacity to do so.

The path forward is not clear. Issues are complex, ambiguous, and morphing; opinions about what is happening and what should be done are becoming polarized and politicized. The context is changing and threatening, and the capacity of human beings and companies to think intelligently is consequently challenged.

Inasmuch as the principal actors feel threatened, they may become closed and rigid in their thinking at the very time in history that we need them to be open and flexible. Inasmuch as coalitions form on alternative sides of major issues, the collective ability to create and implement integrative solutions may be undermined. Inasmuch as the structure of the industry and of society is anchored to an unsustainable global solution, the industry itself may become petrified and unable to address concerns.

From the vantage point of research on organizations that excel in complex, ambiguous, and changing—threatening—contexts we argue that companies and industry

- focus on a truly inspiring goal, that of, *creating an environmentally sustainable world*—a goal that is positive, achievable, meaningful, and urgent—and which has the potential to become a rallying point that unleashes expansive thinking, contribution, and cooperation.
- embed thinking about sustainability in their mental and action models and into their decision-making systems by developing three competencies:
  o "Heedful and Resilient Thinking Focused on Sustainability"
  o "Resilient, High Reliability Interaction, and Response," and
  o "Intercompany Collaboration Toward an Inspiring Goal."

We recognize the pivotal role of inspired and inspiring leadership. And we argue that society should invest in the thinking and acting capacity of industry and cooperate with industry in other ways.

Cooperation can be difficult to achieve in the best of circumstances. We recognize, however, that there are excellent reasons to believe that we will succeed in working together and creating an environmentally sustainable planet. There is widespread recognition that something must be done.

Society is focusing on the issue and acknowledging that it is urgent. Companies and the oil and gas industry are taking action. There are good—well-intentioned, bright, competent—people in each of the groups that are involved in addressing the "sustainability issue" from a variety of perspectives and with a variety of strategies, many of whom are connected to one another in identity-based networks that foster cooperation (Milton and Westphal 2005).

The way forward is for "us" to recognize and build upon the factors that are hopeful and to use the energy that we gain from this to intelligently (and realistically) minimize and work around the factors that could undermine us. This may seem simplistic. It may seem obvious. What is less so is what we need to do. What is less so is how fear and pessimism may undermine all concerned at a time when we need to work together.

Rather than lamenting on what can undermine us, we decided that we would rather unite our science and industry perspectives of the situation we are in—positioned largely with the oil and gas industry on one side and society on the other—to illuminate a path forward. In so doing we have shared what it is that gives us hope that there is a way out of the box we are constructing, what it is that gives us pause, and what we see as a way forward. Our aim has been to contribute to a conversation about where we go from here—what the oil and gas industry and society and their respective institutions may be able to do now to move forward. As a starting point, we focus on what science suggests about developing the critical thinking and acting capacity of the oil and gas industry as a step—one step—toward a more environmentally sustainable industry and world.

Acknowledgments

We thank George Brindle for sharing his knowledge of the energy industry with us. His contribution helped us to appreciate the effort that the oil and gas industry is already making to become environmentally sustainable and the excellence it is achieving on this path. His constant questioning and insights helped us to carefully consider, and often reassess, our science perspective on what steps industry can now take. The constructive comments of Brian Quickfall similarly helped us to develop our perspective on how the multiple actors focused on environmental sustainability can work together. We recognize and thank Zhiwei Han, Katrina Montgomery, and Brittany Harker Martin for their library research in support of us writing this chapter. We thank Jenny Hoops for her editing and for her reflective mind that helped us to refine our argument. Thinking in tandem with those who helped us in crafting this chapter reminded us that science, like industry, is a social enterprise. We must all work together to create a world within which life and industry can thrive—a world worth passing on to our children.

References

Argyle, M. 1991. *Cooperation: The basis of sociability.* London: Routledge.
Axelrod, R. 1984. *The evolution of cooperation.* New York, NY: Basic Books.
Beyer, J. M., and Milton, L. P. 1994. *Being a SEMATECH Assignee.* SEMATEC (internal publication).
BG Group. 2007. *BG Group Corporate Responsibility Report 2006.* London: BG Group.
BP p.l.c. 2006. *BP Sustainability Report 2006.* London: BP p.l.c.
BP p.l.c. 2007. BP Statistical Review of World Energy June 2007. London: BP p.l.c.
Blair, T. 2007. *Global Relations: A Conversation with Tony Blair.* Speech presented October 26, 2007, Telus Convention Centre, Calgary, Alberta.
Bleeke, J., and Ernst, D. 1991. The way to win in cross-border alliances. *Harvard Business Review* 69 (6): 127–135.
Bradsher, K., and Barboza, D. 2006. Pollution from Chinese coal casts shadow around globe; the energy challenge: the cost of coal. *New York Times.* June 11, 2006: 1.1. Late edition, East Coast.
Brown, S. L., and Eisenhardt, K. M. 1998. *Competing on the edge: Strategy as structured chaos.* Boston, MA: Harvard Business School Press.
Browning, L. D., Beyer, J. M., and Shetler, J. C. 1995. Building cooperation in a competitive industry: Sematechnthe semiconductor industry. *Academy of Management Journal* 38 (1): 113–151.
Casey, M. 2007. *As pollution spreads from powerplants, China and the world find themselves choking on coal.* Associated Press, November 4, 2007.
Central Intelligence Agency. (2007, 10 18). *CIA—The World Factbook—World.* CIA Website—Central Intelligence Agency: https://www.cia.gov/library/publications/the-world-factbook/print/xx.html (accessed October 27, 2007).
Chevron Corporation. 2007. *2006 Corporate Resonsibility Report.* San Ramon: Chevron Corporation.
Dallaire, R. 2004. *Building sustainable value.* Guest Speakers Series—On Ethical Leadership. Richard Ivey School of Business. London, ON: February 6, 2004.
Dodgson, M. 1993. Learning, trust and technological collaboration. *Human Relations,* 46: 77–95.
Druskat, V. U., and Wolff, S. B. 2001. Building the emotional intelligence of groups. *Harvard Business Review* 13 (5): 80–90.
Dutton, J. E., Dukerich, J. M., and Harquail, C. V. 1994. Organizational images and member identification. *Administrative Science Quarterly* 39: 239–263.
Environment Canada. 2007. *Acid Rain and the Facts.* Government of Canada, Environment Canada Web Site: www.ec.gc.ca/acidrain/acidfact.html (accessed October 15, 2007).

Foreign Affairs and International Trade Canada. 2007. *Key Documents on the History of Canada's International Affairs.* Government of Canada, Foreign Affairs and International Trade website. http://www.dfait-maeci.gc.ca/department/history/keydocs/menu-en.asp (accessed October 30, 2007).

Ferraro, F., Pfeffer, J., and Sutton, R. 2005. Economics language and assumptions: how theories can become self-fulfilling. *Academy of Management Review* 30 (1): 8–24.

Fredrickson, J. W., and Mitchell, T. R. 1984. Strategic Decision Processes: Comprehensiveness and Performance in an Industry with an Unstable Environment. *Academy of Management Journal* 27: 399–423.

Golembieski, R. T. 1989. *Organization development: Ideas and issues.* New Brunswick, NJ: Transaction Publishers.

Gray, B., and Wood, D. J. 1991. Collaborative alliances: Moving from practice to theory. *Journal of Applied Behavioral Science* 27: 3–22.

Grayson, D. and Hodges, A. 2004. *Corporate social opportunity!: Seven Steps to make corporate social responsibility work for your business.* Sheffield, England. Greenleaf Publishing.

Gulati, R. and Nohria, N. 1992. Mutually assured alliances. In J. L. Wall, and L. R, Jauch (Eds.). *Academy of Management Best Paper Proceedings:* 17–21.

Huber, G. P. 2004. *The necessary nature of future firms: Attributes of survivors in a changing world.* Thousand Oaks, CA: Sage Publications.

Kelly, J. R., Futoran, G. C., and McGrath, J. E. 1990. Capacity and capability: Seven studies of entrainment of task performance rates. *Small Group Research.* 21: 283–314.

Lindblom, C. E. 1959. The science of muddling through. *Public Administrative Review* 19 (2): 78–88.

Locke, E., and Latham, G. 1990. *A theory of goal setting and task performance.* Englewood Cliffs, N.J.: Prentice-Hall.

Margolis, J. D., and Walsh, J. P. 2003. Misery loves companies: Rethinking social initiatives by business. *Administrative Science Quarterly* 48: 268–305.

Masten, A. S., and Reed, M. J. 2002. Resilience in development. In *Handbook of positive psychology,* ed. C. R. Snyder and S. J. Lopez, 74–88. New York, NY: Oxford University Press.

McKibben, B. 1989. *The end of nature.* New York, NY: Random House.

Meadows, D. H., Meadows, D. L., Randers, J. and Behrens III, W. W. 1972. *The limits to growth.* New York, NY: Universe Books.

Meadows, D., Randers, J., and Meadows, D. 2004. *Limits to growth: The 30-year update.* White River Junction, Vt: Chelsea Green Publishers.

Milton, L. P., and Milton, A. P. 2000. Managing diversity strategically in a project-based world. In *Proceedings: Fourth international conference of the International Research Network on Organizing by Projects.* Sydney, Australia.

Milton, L. P., and Westphal, J. D. 2005. Webs of support: Identity confirmation networks & cooperation in work groups. *Academy of Management Journal* 48: 191–212.

Milton, L. P. 2007. Educating for a sustainable world. *Invited Presentation, Fordham University.* New York City, NY. May 9, 2009.

Oliver, C. 1990. Determinants of interorganizational relationship: Integration and future directions. *Academy of Management Review* 15: 241–265.

Orlitzky, M., Schmidt, F. L., and Rynes, S. L. 2003. Corporate social and financial performance: A meta-analysis. *Organization Studies* 24 (3): 403–441.

Royal Dutch Shell plc. (2007). *The Shell Sustainability Report 2006.* Bankside: Royal Dutch Shell plc.

Shabecoff, P. 2001. *Earth rising: American environmentalism in the 21st century.* Washington, DC: Island Press.

Sharfman, M. P., Gray, B., and Yan, A. 1991. The context of interorganizational collaboration in the garment industry: An institutional perspective. *Journal of Applied Behavioral Science* 27: 181–208.

Sitkin, S. B. 1992. Learning through failure: The strategy of small losses. *Research in Organization Behavior* 14: 231–266.

Statoil ASA. (2006). *Statoil and sustainable development 2005.* Stavanger: Statoil ASA.

Staw, B. M., Sandelands, L. E., and Dutton, J. E. 1981. Threat-rigidity effects in organizational behavior: A multilevel analysis, *Administrative Science Quarterly* 26 (4): 501–524.

Stoner, J. A. F. 2006. *It's not about the profits.* Keynote address: Third International Research Conference on Business Management: 2006, "Corporate Social Responsibility: Profits and Beyond"; University of Sri Jayewardenepura, Gangodawila, Nugegoda, Sri Lanka. March, 2006.

Stoner, J. A. F., and Werner, F. M. 2006. *Designing organizations for global sustainability: An inquiry into organizational possibility,* Virtual Paper for the Business as an Agent of World Benefit Conference. Cleveland, OH. October 2006.

Suncor Energy. 2007. *Climate change. A decade of taking action.* Calgary, Ab: Suncor Energy.

Sutcliffe, K., and Vogus, T. 2003. Organizing for resilience. In *Positive organization scholarship,* ed. K. Cameron, J. Dutton and R. Quinn. San Francisco, CA: Berrett-Koehler.

Tapscott, D., and Williams, A. D. 2006. *Wikinomics: How mass collaboration changes everything.* London, England: The Penguin Group.

Tertzakian, P. 2007. *A thousand barrels a second: The coming oil break point and the challenges facing an energy dependent world.* New York, NY: McGraw-Hill.

The Nobel Foundation. 2007. *Norwegian Nobel Committee Press Release.* October 12, 2007. Oslo, Norway.

Thomas-Hunt, M. C., and Gruenfeld, D. H. 1998. A foot in two worlds: The participation of demographic boundary spanners in work groups. *Research on Managing Groups and Teams,* 1: 39–57.

Trice, H. M. and Beyer, J. M. 1988. Using six organizational rites to change culture. In *Gaining control of the corporate culture,* ed. R. H. Kilmann, M. J. Saxton, and R. Serpa, 370–399. San Francisco, CA: Jossey-Bass.

United Nations. 1987. *Our common future.* World Commission on Environment and Development (1987). New York, NY: Oxford University Press.

Urdal, B. T. 2007. *Sustainability leader.* Zurich: SAM Research AG.

Wanberg, C. R., and Banas, J. T. 2000. Predictors and outcomes of openness to changes in a reorganizing workplace. *Journal of Applied Psychology* 85: 132–142.

Weick, K. E., and Roberts, K. H. 1993. Collective mind in organizations: Heedful interrelating on flight decks. *Administrative Science Quarterly* 38: 357–381.

Wernerfelt, B. 1984. A resource-based view of the firm. *Strategic Management Journal* 5: 171–180.

Willard, B. 2002. *The sustainability advantage: Seven business case benefits of a triple bottom line.* Gabriola Island, BC: New Society Publishers.

Wood, R. E. 1986. Task complexity: Definition of the construct. *Organization Behavior and Human Decision Processes* 37: 60–82.

PART IV

Sowing the Seeds for a Viable Future

CHAPTER 8

Designing Undergraduate Education on "Managing for Sustainability"

Paul Shrivastava, Douglas E. Allen,
and Tammy Bunn Hiller

The Context for Global Sustainability Education

A number of high-profile media stories in recent years have made social and environmental issues prominent in the minds of the American public and corporations. Nobel Laureate Al Gore's Academy Award–winning documentary film on global warming, *An Inconvenient Truth,* and cover stories on social and environmental challenges facing the world economy in *Newsweek, Time, BusinessWeek, The Economist,* and *Fortune* have helped raise awareness of significant problems we have accumulated over the past hundred years of industrial development and the need for ecologically and socially sustainable economic development in the future.

At least for the past decade, management educators have been trying to understand the implications of ecological sustainability for management and organizations. A leading academic association of management professors, the Academy of Management, has established the Organizations and the Natural Environment Division, and numerous journals—such as *Organization and Environment, Business Strategy and the Environment,* and *Journal of Sustainable Strategic Management*—focus on new research in this area.

Concerns about the impacts of economic pursuits on cultures, societies, and communities have a somewhat longer history among management academics. The Academy of Management's Social Issues in Management Division dates back to the early 1970s, and the field is served by such publications as *Business Ethics Quarterly, Journal of Business Ethics, Business & Society,* and *Business and Society Review.*

Although we know of few universities offering a course of undergraduate study specifically in "social issues in management" or "business and society," some universities are trying to educate managers more fully about ecological issues. Some of these offer concentrations in "Environmental Management" or offer joint degrees between the School of Management or Business and the School of Natural Resources or Forestry. Most of these efforts are small, new, and experimental. They try to cross-pollinate management concepts with ecological considerations by having management students take environmentally oriented courses.

Bucknell is one of very few universities working to build a program on sustainable management from the ground up, starting from the concept of "sustainability," identifying sustainability management concepts, skills, and tools and developing an educational program to deliver them. Such a project has the unique advantage of starting with a clean, substantive, and ideological slate and creating program philosophy, content, and courses specifically suited to sustainable management. It also avoids the tendency common in current programs of simply adding sustainability as a topic, course, or set of courses to a traditional management program. In this paper we describe our efforts to develop undergraduate education in "Managing for Sustainability (MfS)" in the context of a liberal arts university.

For many years, Bucknell's Management Department has prided itself in teaching management as a liberal art. As a recent department curriculum-review document noted:

> Truly professional education is, by necessity, liberal. The habits of thought associated with liberal education—free inquiry, moral reasoning, engagement with traditions of knowledge and culture, and critical thinking—are precisely the qualities most required of true professionals in all areas. Engineers, teachers, managers, and even musicians, though they require specialized knowledge to succeed in their domains, are only professionals insofar as they understand their endeavors in relation to society's needs, the forces of history, the bounds of responsible practice, and the nature of the human condition. At its best, management education helps students to understand organizations and their management in just such a context.

What we report here is still a work in progress and stems from the efforts of our department's curriculum-review process, which has been in progress for over one year. One aspect of this review was to conceptualize and develop curricula in five areas, of which "MfS" is one. The paper begins by reflecting on and synthesizing our understanding of the concept of sustainability and sustainable development, which has its roots in environmental concerns but which also incorporates, of necessity, concerns about social issues as they are related to economic pursuits. Then we describe a program proposal on Managing for Sustainability with core and elective courses. We close with some challenges and limitations faced in educating for sustainability.

Evolution of the Sustainable Management Concept

Management-sustainability education must begin with a sociohistorical understanding of ecological sustainability. The most recognized definition of sustainability stems from the World Commission on Environment and Development (WCED) and its landmark 1987 *Brundtland Report* (World Commission on Environment and Development 1987; also entitled *Our Common Future*), which focused and energized the sustainable-development movement. The report describes sustainable development as "development that meets the needs of the present without compromising the ability of future generations to meet their own needs." The elegant parsimony of this description masks the underlying complex, heterogeneous, and sometimes contradictory character of this paradigm and its evolution. The use of both *sustainability* and *sustainable development* in the report, and the distinction between them, is a reflection of important contested ground underlying the development of this concept. The growth of the environmental movement during the nineteenth and twentieth centuries places into sociohistorical context current understandings of the modern sustainable-development concept (Edwards 2005). In the brief review below, we chronicle the early environmentalist movements and their impact on the current idea of sustainable development. This history reveals the competing views surrounding the relationship between economic growth and the environment, as well as the reconciliation of these seemingly antithetical interests. The impulses underlying sustainable development have now evolved to embrace many aspects of our economy, organizations, and corporate culture and practices.

While consideration and concern for the environment dates back to ancient Greece and biblical times, the Transcendentalist movement of the nineteenth century was the modern source of the sustainability-movement paradigm. Influenced by Romanticism and mystical spiritualism, Transcendentalists such as Emerson and Thoreau viewed nature as reflecting divinity. This stance culminated in the "preservationist" strain of environmentalism, advocating that parts of the natural environment should be sheltered from all human intrusion whatsoever. At the same time a competing environmentalist position, termed the "conservationist" movement, developed, also advocating that natural areas should be protected. However, unlike the preservationists who viewed the protection of nature as a value in itself, the conservationists advocated protection of nature so that it could be harnessed for subsequent use and enjoyment of people (Robinson 2004). Though subtle, the distinction between the preservationist (ecocentric) and conservationist (anthropocentric) positions reflects a fundamental enduring divide over whether nature should be treated as sacrosanct or used instrumentally in serving the interests of humans.

This dichotomy set the stage for further nuanced alternative views of the environment during the twentieth century. The preservationist view emphasizes that environmental degradation problems should be addressed by fundamental changes in modern cultural values and perspectives, such that they are more in accordance with the preservation of nature. The conservationist perspective places emphasis on the belief that nature can be harnessed or managed and that technology and collective policies are the solution to problems with the environment. Thus, the preservationist and conservationist schools of thought differ in the degree to which concern for the natural environment is at odds with modernity's fundamental commitment to economic growth. These earlier contrasting logics can be used to understand the rationale underlying sensitivity to the usage of the terms "sustainability" and "sustainable development." Paralleling the preservationist school, the term "sustainability" is preferred by some who believe that sustainable development is an oxymoron. That is, instead of altering the nature of development in the direction of greater sustainability, what is needed is the fundamental realignment of cultural values and practices associated with development (e.g., accumulation and mass consumption). Alternatively, conservationists adopt the term "sustainable development" and place more emphasis on changing technological processes to accommodate growth and development in a manner compatible with the environment.

In this light, the release of the Brundtland Report was a watershed event in that it set the stage for a sustainable development paradigm by adopting as axiomatic the assumption that concern for the environment and concern for economic development do indeed go well together. In fact, not only did the Brundtland Report accept the compatibility of environmental concerns and economic development, it viewed growth in economic activity as essential. The reason for this seemingly counterintuitive conclusion is that adoption of the sustainable-development perspective enabled the commission to reconcile another dispute that pitted environmental advocates against those concerned about human development and poverty in less-developed areas of the world. While environmentalists' greatest concern centered on overdevelopment, advocates for impoverished people and countries around the world were mostly concerned about underdevelopment. The Brundtland

Report represents a compromise among these three competing concerns— economic, environmental, and human welfare. Its adoption of the terminology of sustainable development and its underlying assumptions seem to have substantially impacted the contemporary scope of the movement. Testament to this impact can be found in the fact that until very recently the terms "sustainability" and "sustainable development" seem to have been used interchangeably by most people and organizations currently working in this paradigm.

The "three E's" (ecology/environment, economy/employment, and equity/ equality) most often associated with sustainable development are a straightforward extension of the synthesis of concerns formulated in the WCED (Edwards 2005). Perhaps the most prominent dimension readily associated with sustainability is its focus on the environment and ecology. Of course, the sustainability movement is differentiated from its roots in earlier ecological and environmental movements given its explicit assertion that economic development and employment opportunities are not completely antithetical to environmental concerns. Sustainability also extends to social concerns over how environmentally sensitive development can be managed to enhance global equity and equality of material well-being. Though less prominent, sustainability also incorporates a concern for how economic development can unfold in a way that is respectful of and preserves long-standing diverse cultures, that fosters an environment at human scale, and that is conducive to politically stable democracies.

Corporate Incorporation of Sustainable Development

In part because the concept of sustainable development has an incredibly broad scope, encompassing the confluence of environmental, economic, and social concerns, many have criticized it since its introduction. Given its potential to be interpreted in many ways, some assert that it has no real meaning. Other skeptics have characterized the concept as contradictory or vague (Daly 1996; Dovers and Handmer 1993). These criticisms notwithstanding, no one debates that many of the impulses underlying the sustainable-development paradigm, sometimes under the banners of corporate social responsibility or corporate environmental responsibility, are affecting corporations and their practices.

Corporations' relationships with the themes underlying sustainable development have been fluid and dynamic (Hoffman 1997). Until the mid-1960s, U.S. corporations operated in a cultural climate that foregrounded the salutary impacts of corporate productivity, while leaving in the shadows concerns about environmental degradation. The playing field began to change in response to events such as the publication of Rachel Carson's influential book *Silent Spring* (1962). In the 1970s government regulations began to hold corporations more accountable for environmental impacts. Though most corporations' actions to mitigate their environmental impacts were motivated instrumentally by the rationale of avoiding legal penalties, environmental concerns began to appear on their radar screens. The election of Ronald Regan led to an era during which the government mediation of the conflict between environmentalists and corporations waned, leading to direct confrontation. In response, corporations more directly addressed environmental issues and began to categorize environmental concerns with other social responsibilities. By the late 1980s, environmental concern had become engrained in corporate policies. Before long, insurance companies, customers, and investors began to be seen as advocates for explicit environmental stakes. Some corporations even began considering environmental outcomes in the context of their corporate strategies.

Although a historical perspective may demonstrate the evolution of corporate environmentalism at the institutional level, the contemporary imperative for corporations to embrace sustainable development is driven by recent circumstances. Following Gusfield (1992), social movements such as sustainable development have enduring historical roots that persist over time while also adapting to the unique circumstances characterizing particular epochs. For example, the roots of the natural- and organic-foods movement can be traced back to the 1830s and represent a cultural backlash to modernity. Natural foods were viewed as healthier because they were the antithesis of modern food production and were demarcated in opposition to modernity by their lack of refining, closer proximity to the end consumer, and smaller scale. Unlike its earlier strain under which natural foods were preferable for their health benefits, in the 1960s, natural and organic foods began to be embraced for their value to the environment. Similarly, many of the movements subsumed in the sustainable-development paradigm (e.g., environmentalism, development economics, community-supported agriculture, etc.) arguably have their roots in resistance to modernity. However, as Gusfield points out, the cultural themes underlying movements are not exactly reproduced over and over again across time. Rather, they adapt to the unique circumstances of a particular epoch. As such, the current incarnation of the various movements subsumed under the aegis of sustainable development may be better grasped as a reaction of cultural resistance to late-/postmodern societal conditions.

In late/postmodern capitalism, economic activity is characterized by globalization, such that corporate actions transcend both the national boundaries and national regulatory bodies. Investment capital can flow freely and rapidly from one region of the world to another. Corporations can shift operations somewhat quickly from region to region based on the cheapest labor rates. Global supply chains can transport food from distant locales with consumers having little information about its origin or the conditions in which it was grown. In short, the qualities of late/postmodern capitalism extend corporations' tremendous powers.

At the same time, some of the very same qualities of postmodern society that have empowered corporations, such as those related to technology, transportation, and communication, have also empowered other organizations concerned about sustainable development to resist corporate impacts. For example, in the 1990s, NGOs such as Rainforest Action Network and Greenpeace joined with local indigenous peoples to stop MacMillen Bloedel from logging the old growth rainforest in Clayoquot Sound, British Columbia. The activists enlisted the support of MacMillen Bloedel customers, such as News International, Kimberly Clark, and Scott Paper, who joined the effort because they feared that their own customers—end-user consumers—would boycott products made from old growth rainforest.

The general point is that corporations have now assimilated not only environmental concerns but related social concerns as well, just as one might have anticipated based on a historical examination. Interactions between corporations and environmentalists are no longer mediated by governmental agencies. In a shrinking world marked by closer proximity of all the world's actors, it stands to reason that the sustainable-development paradigm is growing deeper roots.

While this is a welcome development, we acknowledge that corporate practices in this area are rather limited and tentative. Corporate environmental efforts today focus largely on technical operations, such as reducing the use of virgin materials, using ecologically efficient production methods, pollution prevention, ecofriendly product design and packaging, and waste management. Companies are adopting environmental-management programs that are technologically feasible and save costs (harvesting the low-hanging fruit). Many companies are also making deeper systemic changes by adopting ISO 14000 environmental-management systems (Hillary 2000). The World Business Council for Sustainable Development (www.wbcsd.org) lists numerous ecological sustainability programs adopted by the world's largest corporations (Sharma and Starik 2002; Shrivastava 1996).

Yet these corporate changes may be best characterised as incremental and piece-meal. Companies cautiously engage environmental solutions that have clear financial benefits. Some undertake environmental projects for their publicity value in the hopes of gaining legitimacy and attracting new customers. On a broader scale, companies are still reluctant to make large-scale risky investments in sustainability ventures that would fundamentally transform corporate strategies and operations (Roome 1998).

Sustainability Management Education

Organizations of all types are acknowledging the need to understand basic issues of sustainability at both global and local levels. Large companies such as Wal-Mart and Proctor & Gamble have created executive vice president–level positions and departments to incorporate sustainability at the strategic level. The World Business Council for Sustainable Development (a group of over 1,000 large corporations) developed "Sustainable Development: A Business Primer" as a training tool that laid out basic ecosystem concepts that all managers should know. But these corporate groups also see that sustainability is not just about improving ecological performance while maintaining financial health: they also address ethical, social-justice, human-rights, equity, and security issues. Sustainability in the corporate management context is thus a multifaceted variable.

In modern societies virtually every aspect of life (economy, politics, culture, etc.) is influenced or controlled by organizations. Organizations are the primary engines for the creation of wealth and other services of social and cultural value. Their size and scope of activities make them a vital part of the societal infrastructure and national and international cultures. They help shape and reproduce those cultures. Organizations' system-wide impacts and roles require that we take a systemic view of organizational sustainability to include not only economic and ecological performance but also social and ethical performance.

Savitz and Weber (2006) addressed the concept of the "triple bottom line," as one way of understanding corporate sustainability: sustainable corporations are concerned with their environmental and social outcomes, as well as their financial outcomes. Sustainable corporations create profit for their shareholders while acting to foster the well-being of society and the protection of the environment. Every subsystem of these organizations—from strategy to day-to-day operations—incorporates regular stakeholder engagement to build trust and open communication. Companies also develop multidimensional measurement and reporting systems to assure accountability and transparency.

All organizations—for-profits, NGOs, and governmental bodies—can benefit from a triple-bottom-line approach, which makes it an ideal way to frame management education. At Bucknell, we acknowledge that corporations must earn a reasonable return for shareholders in order to advance, but we are also committed to helping our undergraduate students learn to balance these goals with concerns for social and environmental justice. After all, the long-term survivability and health of humans are dependent upon our ability to live with one another and within the bounds of the natural environment. Corporate financial success is hollow when it is constructed through exploiting today's poverty-stricken and desperate peoples; and leaving future generations of mankind with a degraded environment that is no longer able to provide adequately for them is unconscionable. In order to build a world that we can all be proud of, today's managers must begin to give greater consideration to the environmental and sociocultural impacts of their decisions. They must adopt new priorities, encourage innovative thinking of all types within their organizations, and seek alternative means for achieving success.

A Rationale for Undergraduate Education in Managing for Sustainability

Why should undergraduates study sustainability? What is so different or special about managing for sustainability that it deserves the focused attention of aspiring managers? In this section we present several important rationales for choosing to include sustainability management at Bucknell. Our overall rationale is that the course, Managing for Sustainability, offers a holistic and ethically complete approach to dealing with urgent managerial challenges in a global economy. Let us parse this rationale out in some more detail.

Integrative holistic understanding of management issues

In the past, undergraduate management curricula have focused largely on developing students' understandings of the functional aspects of organizations. Less attention has been paid to the social, historical, and ecological contexts in which managing and organizing happen. At Bucknell, in our attempt to teach management as a liberal art, this contextualized understanding is critical. Our students are relatively young; they have little or no work experience or exposure to corporate life to appreciate the realities of managing organizations. To make management issues real and meaningful to them, they need a holistic, contextualized understanding of management that includes social and ecological components as well as economic ones. MfS, as we have conceptualized it above, provides a larger canvas for discussing traditional management topics such as leadership, strategy, organizational behavior, organizational design, and human-resources management, in the light of ecological, social, and economic challenges facing the world.

Globalizing economies and globalizing ecological concerns

Managers and policy makers are beginning to realize that globalization of the economy is an inexorable trend. In the next fifty years the global economy is expected to expand to five to ten times its current size. Unfortunately, one unintended consequence of expanding economies is large-scale destruction of ecosystems. From a global perspective the production of wealth and goods is fundamentally tied up with the consumption of ecological assets. Our graduates, whose managerial choices will impact not only their companies but the environment as well, need to understand the reciprocal relationship between production systems and ecosystems. The managing of corporations cannot be studied in isolation from the ecological impacts of corporations, so managing for sustainability is simply a prudent and progressive orientation for all managers.

The ethical imperative

Inherent in the functions of management is a set of societal expectations and ethical and legal responsibilities. As our understanding of organizational impacts on society and ecosystems deepens, social and ethical responsibilities change and expand. Management education should prepare managers for the continually evolving ethical imperative of management. Managing for Sustainability moves and relocates ethical concerns in management from the periphery to the center. It acknowledges the variety of stakeholders served by organizations. It recognizes the natural conflicts that exist among them as well as the larger conflicts between the North and the South, the current generation and future generations, and investors and society at large. It offers a framework for debating and resolving the conflicts of interests inherent in managing. It offers opportunities for an open discussion of management values and human values. An education in managing for sustainability can provide students with knowledge about and understanding of the world as it exists, but also help them determine how to structure their own personal values, which will ultimately shape that world. In order to manage for sustainability, our students will understand the need to seek an economic, social, and ecological balance to enable the survival of humans in harmony with nature.

Practical Urgency and Continuity

Many ecological problems that need our attention urgently are rooted in past mismanagement or incorrect assumptions by government and private corporations. For over a quarter of a century, the scientific community has been addressing systemic problems such as ozone depletion, global warming, species extinction, and irreversible changes in aquatic, mountain, and desert ecosystems; yet governments and businesses have not taken adequate action. As a result, what were once relatively circumscribed problems have reached huge proportions and a global scale. Even if we begin now we may be unable to resolve these problems in time to prevent irreversible damages, so there is a real practical urgency for managers to understand ecologically and socially sustainable management.

A sense of practical urgency is already being inculcated in students at the elementary- and secondary-school levels. Students are educated in environmental science and urged to participate in civic-environmental programs such as recycling. They arrive at college with an ecological awareness and expectation of continuity in developing their ecological knowledge base. A management-for-sustainability program can provide both continuity and an extension of this knowledge.

Career Opportunities

The Japanese Ministry of International Trade and Industry forecast that by 2050 over half of global GDP would come from energy and environmental sectors. This GDP forecast included many forms of renewable-energy technologies, niche environmental services, waste management, and pollution-control equipment. Concomitant with this sort of economy will be a set of customer demands, business relationships, and managerial skill expectations. What are the career implications of this forecast for undergraduate management education? While it is difficult to estimate how many environmental engineers and environmental managers would be needed to support this type of economic production, we can safely say that *all* engineers and managers will need at least some basic knowledge of environmental issues to be successful in such an ecology-integrated economy. They will need to understand ecological consequences of work practices and worker behaviors. Undergraduate management education would do well to build ecological awareness and sensibilities in students to make them understand the interrelatedness of economic and ecological systems. One may be employed as an accountant or electronic engineer, but accountants and engineers with an ecological perspective will be able to add more value to their jobs.

Managing for Sustainability (MfS) Program at Bucknell

The context of our curriculum design is undergraduate management education at the College of Arts and Sciences at Bucknell University. Bucknell is the largest national liberal arts college, with an enrollment of about 3,400 undergraduates and fifty-four majors. It offers a Bachelor of Science in

Business Administration (BSBA) through its Management Department located within the College of Arts and Sciences. Bucknell's undergraduate curriculum consists of thirty-two one-credit courses. Under the Common Learning Agenda all students in the College of Arts and Sciences take between eleven and sixteen courses that are distributed across the natural sciences, humanities, and social sciences. The remaining courses are from the student's major and minor areas and electives.

The Management Department's curriculum review conceptualized the BSBA degree in terms of a core curriculum and a choice of five specialized major programs. The core curriculum cultivates three forms of literacy (foundational, managerial, and integrative) relevant to managerial thought. Foundational Literacy helps students discover connections between the study of management and the fundamental disciplines from which it derives: economics, mathematics, the behavioral sciences, and the humanities. Managerial Literacy teaches students how to apply disciplinary knowledge in solving managerial problems. And Integrative Literacy teaches students how to integrate knowledge from multiple perspectives in addressing complex, multifaceted organizational problems.

In addition to the core curriculum, students select a major from among the specialized programs including Accounting and Financial Management; Global Management; Managing for Sustainability; Managing Information and Technology; and Marketing, Consumption, and Innovation. Each of these interdisciplinary programs spans the functional disciplines of management and builds bridges between the arts and sciences and engineering curricula across the university.

Managing for Sustainability is a transdisciplinary program that requires collaboration across natural sciences, social sciences, humanities, and professional disciplines. It integrates management learning with the liberal arts, and, at the same time, it contextualizes that education—both socially and historically.

The curriculum for the Managing for Sustainability program has been under discussion in the department for over a year. It has not been finalized, but our current thinking illustrates key aspects of the integrative holistic approach that the Management Department is pursuing. The Managing for Sustainability program will engage students in an interdisciplinary examination of the challenges of managing organizations in a socially, ecologically, and economically sustainable manner. The program is intended to foster students' critical thinking about organizational values and goals. Students will consider how diverse organizations—for-profits, NGOs, and governmental bodies—can be designed and managed to participate effectively in the global economy while simultaneously reducing poverty, hunger, and other manifestations of human inequality; preserving cultural values and community identity; protecting, conserving, and restoring the environment; and upholding the inherent dignity of humans, nonhumans, and ecosystems affected by organizational activities. Students should develop a deep understanding of the historical and scientific basis of our societal and ecological condition and gain core management skills for resolving economic, social, and ecological challenges and building sustainable organizations. Our hope is that graduates from this program will become managers with a deep environmental and social-justice ethos who can redirect current organizational models toward social, ecological, and financial sustainability.

All students pursuing a BSBA degree will take a set of core courses that will serve as a platform and common framework to understand management issues and learn the basic tools for managerial analysis. These include (a) Management: Past, Present, and Future, (b) Introduction to Organization and Management, (c) Managerial Statistics, (d) Economic Principles and Problems, (e) Accounting, (f) Finance, (g) Marketing, (h) Information Systems, (i) Business, Society, and Ethics, and (j) Management Strategy.

Students in the MfS program will take one course from each of the following four categories and two free electives chosen from any of the four categories:

The Legal/Political Framework of Managing for Sustainability—course examples include the Legal Framework of Managing for Sustainability; Political Economy of Global Resources; Political Geography; Introduction to Public Policy; American Public Policy; State and Local Internship Program; Power, Protest, and Political Change; Global Governance

The Cultural/Societal Framework of Managing for Sustainability—course examples include Sustainable Marketing and Consumption; Unemployment and Poverty; Economic Development; Economic Geography; Third World Development; Human Service Systems; The Sociology of Developing Societies; Human Rights; Sociology of Religion

Environmental Sustainability—course examples include Sustainable Energy and Transportation; Sustainable Agriculture; Organizational Challenges of Genetic Commerce; Environmental Ethics; Resources and the Environment; Environmental Engineering; Introduction to Ecological Design; Sustainable Resource Management; History of American Environmental Politics and Policy; Environmental Policy Analysis; Environmental Justice; Environmental Ethics

Human Society Sustainability—course examples include Organizing for Justice and Social Change; Sustainable Human Resources Management; Social Entrepreneurship; Managing with Passion; Crisis and Disaster Management; Anthropology in Action; Global Justice and Social Change; Public Service and Nonprofit Organizations; Community Organizations in Northern Ireland; Field Research in Local Communities; Social Services and Community: A Practicum; Practicing Democracy: Active Citizenship; Community Engagement and Social Change

Students graduating from the Managing for Sustainability program are expected to pursue a diverse range of careers—general management, political advocacy, law, marketing, etc.—in a diverse range of organizations—for-profits, NGOs, and governmental organizations. We would also expect a significant percentage of the students to pursue graduate education in areas related to sustainability.

In Lieu of a Conclusion: Lessons and Challenges

In this paper we argued the need for educating managers to create and manage ecologically, socially, and economically sustainable organizations. Our efforts at Bucknell are still ongoing as the design of the MfS program is refined and sharpened. We don't have any firm conclusions to offer, but we would like to highlight several lessons and challenges we have gathered from the year-long process of discussions and curriculum review.

One of the biggest challenges to advancing sustainability-management education lies in the very nature of the concept of sustainability. It is a multidisciplinary concept that acknowledges the inherent conflicts among the competing concerns of interest groups, classes, nations, and generations. Understanding the intellectual issues at the heart of sustainability requires integrating the natural sciences, social sciences, and humanities. Most faculty are not trained in these various disciplines in any meaningful way, so they do not know how to integrate disciplines and maneuver among disciplinary lines to understand the intricacies of and challenges to organizational sustainability. Moreover, multidisciplinary research does not have the same status as pure disciplinary research, discouraging faculty focused on tenure or promotion from engaging in it. The conflicts inherent in sustainability issues are value laden, subtle, and complex, making them challenging to incorporate into undergraduate coursework.

Another challenge to sustainability education lies in the profinancial and technocratic consciousness and ideologies that undergird management and business education. Taking sustainability seriously requires a social and ecological consciousness and genuine concern for nature and public welfare both locally and globally. Management curricula are dominated by economics, finance, accounting, and technology courses and faculties. There are strong vested interests in place to maintain the status quo. If a course or two on social responsibility or the environment are added to such curricula, it simply serves as a token acknowledgement and does not genuinely encourage students to embrace a paradigm shift regarding organizational relationships with their social and natural environments.

Institutional challenges to bringing sustainability education into any curriculum also seem daunting. Most universities and colleges do not have a focal "department/school of sustainability." Knowledge relevant to sustainability is spread among many different departments. Bringing together faculty from different departments, academic resources, and administrative support to build a sustainability-management program can be a long and time-consuming bureaucratic process.

These challenges of developing and conveying interdisciplinary knowledge, questioning and revising ideological assumptions, and procuring institutional support require significant faculty time, effort, and motivation. Other institutions seeking to develop sustainability education will be well served by recognizing the people-intensiveness of the program-development process and support it with appropriate resources.

References

Carson, R. 1962. *Silent spring.* Boston: Houghton Mifflin.

Daly, H. E. 1996. *Beyond growth: The economics of sustainable development.* Boston, MA: Beacon Press.

Dovers, S. R., and Handmer, J. W. 1993. Contradictions in sustainability. *Environmental Conservation* 20 (3): 217–222.

Edwards, A. R. 2005. *The sustainability revolution: Portrait of a paradigm shift.* Canada: New Society.

Gusfield, J. 1992. Nature's body and the metaphors of food. In *Cultivating differences: Symbolic boundaries and the making of inequality,* ed. M. Lamont and M Fournier, pp. 75–103). Chicago: University of Chicago.

Hillary, R. 2000. *ISO 14001: Case studies and practical experiences.* London: Greanleaf.

Hoffman, A. J. 1997. *From heresy to dogma: An institutional history of corporate environmentalism.* San Francisco: The New Lexington Press.

Robinson, J. 2004. Squaring the circle? Some thoughts on the idea of sustainable development. *Ecological Economics* 48 (4), 369–384.

Roome, N. 1998. *Sustainability strategies for industry: The future of corporate practice.* Washington, D.C.: Island Press.

Savitz, A., and Weber, K. 2006. *The triple bottom line.* New York: John Wiley.

Sharma, S., and Starik, M. (Eds.). 2002. *Research in corporate sustainability: The evolving theory and practice of organisations in the natural environment.* Cheltenham, UK: Edward Elgar.

Shrivastava, P. 1996. *Greening business: Profiting the corporation and the environment.* Cincinnati, OH: Thomson Executive.

World Commission on Environment and Development. (1987). *Our common future.* Oxford: Oxford University.

CHAPTER 9

The Sustainability Coordinator: A Structural Innovation for Managing Sustainability

Gordon Rands, Barbara Ribbens, and David R. Connelly

Introduction

A recent report from the World Wide Fund for Nature, the Global Footprint Network, and the Zoological Society of London estimated that the health of planetary ecosystems, as measured by the populations of over 1300 vertebrate species, declined by about 30 percent from 1970 to 2003. During the same period, the global human ecological footprint—the amount of biologically productive land and ocean required to provide resources used and absorb waste produced—has more than doubled, from approximately 6 billion hectares to over 14 billion hectares. The total human ecological footprint is now estimated to exceed the available carrying capacity by about 25 percent (Hails, Loh, and Goldfinger 2006). The clear trend is toward even greater declines in ecosystem quality and increases in resource use and pollution, thus further exacerbating overshoot beyond the carrying capacity and portending a future overshoot-and-collapse scenario (Meadows, Meadows, and Randers 1992).

These figures demonstrate that environmentally sustainable behavior cannot be viewed as a desirable goal to be worked on in the future, but must be seen as a critical necessity toward which we must strive in the present. In order both to improve the lot of the world's poor and to avoid catastrophic changes in the world that future generations will inherit, we must alter resource consumption and pollution patterns so as to create ecologically sustainable societies. The creation of such societies absolutely requires the widespread creation of ecologically sustainable organizations (Shrivastava 1995; Starik and Rands 1995).

Much of the attention given in the management literature to improving the environmental sustainability of organizations, particularly business enterprises, has focused on innovations in corporate environmental management practices such as pollution prevention, zero emissions, industrial ecology, design for the environment, life cycle analysis, environmental management systems (EMS), and environmental accounting systems. All of these innovations fall within the "systems" element of strategic environmental management (SEM) identified by Starik and Carroll (1992) in their application of the McKinsey 7S framework to environmental management. Six other 7S elements also exist in their development of an SEM framework: skills, shared values, stakeholders, strategies, structure, and style. The role of structure in improving organizational sustainability has received surprisingly little attention.

We believe that such an omission is unwarranted because what appears to be a highly significant structural innovation has begun to emerge over the last decade in organizations striving to become more sustainable. Over one hundred organizations, concentrated largely in the academic sector, have created a new structural role to address sustainability: the position of the sustainability coordinator.

We begin this chapter with an examination of the campus sustainability movement, the context within which this innovation has emerged. Next, we discuss the emergence of this position at colleges and universities, the duties that it entails, and the advantages that it may pose over other structural approaches to managing sustainability. Finally, we examine its potential for spreading to other organizational sectors, particularly that of municipal governments and corporations.

The Campus Sustainability Movement

David Orr (1990, 1992, 1994) has long argued that a critical prerequisite for a sustainable society is environmentally literate students and that a key element in the development of such ecological literacy is that students have the opportunity to engage in hands-on learning about environmental problems at a scale at which they can understand them, recognize the potential to change the situation, and actually participate in creating such change. The ideal scale and locus of such education, Orr suggests, is the students' university itself. Since 1990 a dramatic increase in such attention has occurred at numerous universities, prompted both by faculty desire to educate students more effectively and by student, faculty, and staff desires to decrease the environmental impacts of their universities.

One of the first major efforts occurred at UCLA (University of California, Los Angeles) where a group of graduate students issued a study identifying the degree to which the university was contributing to environmental demands, and proposed that it take specific actions to reduce its impact and serve as a model (Creighton 1998). While the recommendations were not accepted by the university administration, this initiative served as a model for similar actions at numerous other universities. These efforts coalesced into the campus sustainability or greening-the-campus movement.

These sustainability initiatives have been facilitated by the development of a number of institutional mechanisms. In 1990, twenty-three university presidents, chancellors, and provosts from around the world met in Talloires, France, to sign the "Talloires Declaration," committing their campuses to major sustainability efforts. Over 350 institutions of higher education are currently signatories, more than 125 of which are located in the United States and Canada (ULSF 2007).

In 1992 the National Wildlife Federation created its Campus Ecology program to focus on helping students to become involved in addressing campus environmental issues. This program generates advisory materials, conducts periodic surveys of campus greening efforts, issues an environmental report card, provides consulting assistance for students and faculty, and makes funding available for a number of student fellowships to work on greening projects on their campuses (National Wildlife Federation 2007).

In 1994, the Campus Earth Summit, sponsored by the Heinz Family Foundation, took place at Yale University, where students, faculty, and staff from universities in fifty states and twenty-two countries developed the "Blueprint For A Green Campus" (The Heinz Family Foundation 1995). This blueprint presented a set of ten recommendations for greening higher education institutions and suggested strategies for implementing these recommendations. The recommendations address the following areas: improvement and integration of environmental programs and issues, opportunities to study campus environmental resource flows, campus environmental audits, environmentally responsible purchasing, waste reduction, energy efficiency, integration of sustainability approaches into planning for campus facilities and land use, establishment of student environmental centers, and support for students seeking environmental careers.

Since 1996, seven biennial conferences on campus sustainability have been held at Ball State University (BSU). They have provided excellent opportunities for students, staff, and faculty to come together and share their experiences with campus greening efforts (BSU 2007). A similar conference held in the fall of 2004 in Portland, Oregon, led to the creation the following year of the Association for the Advancement of Sustainability in Higher Education (AASHE), which is now the central organization in the campus sustainability movement (AASHE 2007a, 2007b, 2007c). AASHE holds conferences in alternating years with those at BSU, issues a weekly electronic newsletter on campus sustainability happenings, annually bestows Campus Sustainability Leadership Awards on four campuses (a community college, and a small, midsized and large college or university) for their sustainability accomplishments, and provides extensive information resources. AASHE is now developing a variety of initiatives to assist campus sustainability advocates. These initiatives include developing a tracking, assessment, and rating system to help campuses gauge their progress; developing best practices guides; and encouraging campuses to pledge to undertake planned efforts to reduce greenhouse-gas emissions and to become "carbon neutral." AASHE has grown rapidly from a staff of two to a staff of eight in just two years.

In addition to the work of specialized organizations, university-affiliated organizations with broader purposes have begun to devote significant attention to sustainability. One such example is the Society for College and University Planning (SCUP), which has sponsored a Campus Sustainability Day webcast since 2002. The associations representing campus business, facilities, and purchasing officers, among others, have also developed sustainability initiatives. All told, more than twenty organizations have been involved in promoting or contributing to the campus-greening movement in the United States (Second Nature 2007). In Canada, several other organizations have been active in this movement. The Sierra Youth Coalition Sustainable Campus Project now runs an annual conference and has student participants in its program from 75 percent of Canadian universities. The Canadian Consortium on Sustainability Research and the Environmental Studies Association of Canada have also begun to consider campus sustainability issues. In Europe, involved organizations include an organization similar to ULSF (University Leaders for a Sustainable Future), the Copernicus Charter. In addition, the governments of the United Kingdom and the Netherlands have both begun to pay attention to the value and importance of campus greening.

Many universities have conducted campus environmental audits or assessments, and a consulting firm has been established that specifically works with universities to facilitate or conduct such assessments (Skov 2004). A number of different "cross-institutional sustainability assessment tools" have been developed to help identify sustainability best practices, to focus attention on continuous sustainability improvement, and to communicate progress within one campus and to other universities (Shriberg 2002). In addition, a number of universities have begun to implement an EMS, either certifying to ISO 14000 or developing their own (Clarke 2004; Nicolaides

2006; Price 2005). The movement appears to be having a significant impact on facilities operations: a survey found that 84 percent of facilities officers indicated that sustainability is a significant consideration in making a variety of facilities and purchasing decisions (Akel 2006).

The growth of the campus sustainability movement has been impressive and has begun to garner substantial press coverage. Among the topics receiving attention in the national media have been the boom of certified green dormitories on campus (Eastman 2007), official college efforts to encourage green behavior by new students (Benderoff 2007), increased incorporation of sustainability into college curricula (Steptoe 2007), and the commitment of over 400 college and university presidents to have their schools strive for carbon neutrality (Deutsch 2007). In addition, there have been several rankings or assessments of different campuses' sustainability efforts (Grist 2007; Hattam 2007; Sustainable Endowments Institute 2007).

Structural Approaches to Campus Sustainability

With all of these sustainability efforts occurring, how are they being managed? Sustainability efforts usually begin with an individual or small group of champions, either faculty or students, growing into an ad hoc effort (e.g., Bartlett 2004; Uhl 2004). At many campuses this situation is still the state of affairs. A process of formalization is clearly under way, however. Many universities now have a sustainability task force or committees made up of some combination of administrators, staff, faculty, and students that examine or propose—and in many cases are responsible for implementing—changes in campus environmental practices (Starik, Schaeffer, Berman, and Hazelwood 2002). While currently the most common approaches to the management of campus sustainability, these task forces and committees may eventually be seen as a transition stage. Approximately ninety colleges and universities now have full-time sustainability coordinators (Dautremont-Smith 2007). Many of these coordinators are assisted by other full- or part-time staff members or student interns, occasionally formally organized within a campus sustainability office. Many observers of and participants in the campus sustainability movement believe that the hiring of a sustainability coordinator is a necessary step in making truly significant advances toward campus sustainability.

Roles, Duties, and Skills

The roles, duties, and skills that the sustainability coordinator position requires are many and varied. A guide prepared for colleges contemplating hiring a sustainability coordinator lists twenty-seven different roles and duties (Campus Consortium 2006). The following roles and duties were considered to be particularly important in a review of twenty-eight sustainability coordinator job descriptions (AASHE 2007b)—the number of job descriptions mentioning each item is listed in parentheses: ensuring communication (28), designing or implementing sustainability initiatives (27), maintaining public relations (21), administering budgets and grants (17), providing support for or leadership of a sustainability committee (17), conducting research/data analysis (17), educating others regarding sustainability (14), serving as a liaison across the campus (13), managing student interns (11), and assessing and evaluating programs (11).

That list is consistent with the activities and duties described by sustainability coordinators themselves in two focus groups held at the 2005 Greening the Campus conference (Rands, Ribbens, and Bingham 2006). The focus group members described their most important activity to be education and outreach to university students, staff, and faculty. What sustainability means to a university campus, and how to improve it, is not widely understood. Educating and energizing students was described as especially critical, particularly given the continually changing cast of students as they progress through college. Communication with external stakeholders (vendors, the local and state community, and other sustainability coordinators) was also described as being quite important, particularly in terms of communicating needs and challenges that exist and collecting information on practices that can be applied to the campus. Working with the sustainability committee that exists on most campuses was also seen as very important. Coordinators described their roles as involving tremendous amounts of facilitation, in which they provide others with resources including expertise, coordination, and networking assistance. The ability to connect different parties, with similar interests and concerns or possessing complementary resources, from within and outside the campus was regarded by coordinators as especially important. Making these connections was seen as critical to successfully enacting more sustainable practices than would occur without those parties' involvement or if the sustainability coordinators were to be solely responsible for implementing activities.

Most coordinators indicated that they were responsible for managing budgets, as well as for supervising consultants and students—either interns or employees. They also reported being personally responsible for taking action. They develop grant proposals and action recommendations, plan and implement projects, evaluate initiatives and disseminate their results, and often teach courses on organizational sustainability. Finally, they also reported that their duties and activities go well beyond those listed in their formal job descriptions.

The skills needed by sustainability coordinators are many and varied, including the ability to analyze problems and activities using multiple lenses such as environmental, technical, financial, political, and behavioral dimensions. Skills commonly specified in the position descriptions that were examined include communication, leadership, self-management, the ability to understand technical and scientific issues and interact with individuals in those fields, human relations, organizational, computer, strategic, and political skills. In short, sustainability coordinators engage in significant and wide-ranging managerial activity and require many of the same skills as effective managers.

Reporting Relationships

One of the interesting questions regarding the sustainability coordinator position is where it should be located in the organizational structure. The twenty-eight sustainability coordinator job descriptions discussed above and the titles and departmental affiliations of another eighty full- or part-time coordinators for which job descriptions were not available (AASHE 2007c) were examined. Of the 108 positions, 44 (approximately 40 percent) clearly report to the director of physical plant/campus facilities. Other common affiliations include the sustainability committee or office (16, 15 percent); the office of a nonacademic vice president (VP) or dean (13, 12 percent); an environmental health and safety or environmental affairs unit (12, 11 percent); an environmental institute, center, or environmental studies department (9, 8 percent); the office of the president (4, 4 percent), or the provost (2, 2 percent). One coordinator is located in another academic unit, two others report directly to two different offices (provost and VP for administration; provost and physical plant), and for four the affiliation cannot be determined. It is possible that most of the coordinators listed as lodged within an environmental health and safety/environmental affairs unit are in fact also located within the physical plant/facilities structure. This physical plant/facilities reporting arrangement may apply as well to a majority of those listed as being located in a sustainability office or committee. At least two such offices, however, are listed as reporting to the VP of Administration and Finance. It is clear that a relatively small number of coordinators (fourteen, including joint reporting relationships) report to academic rather than nonacademic superiors.

There appears to be a widespread preference for separating sustainability from environmental compliance. This is not universal, however, as twelve of the coordinators are located within environmental health and safety units. At the University of Connecticut, which has five campuses, the staff member in charge of sustainability issues is also in charge of environmental compliance and reports directly to the VP and Chief Operating Officer of the university (Miller, Umashankar, and Mella 2005).

As the predominant focus of the campus sustainability movement has been on reducing the direct environmental footprint of universities, finding sustainability coordinators located within nonacademic units is not surprising. Changes in the use of physical resources such as energy, water, food, and materials are under the control of facilities, housing, dining, and purchasing units, not under the faculty. However, incorporating attention to sustainability in the curriculum is increasingly of interest within the movement, and many sustainability coordinators occasionally or regularly teach a class on campus sustainability issues. While this is easily accomplished on an adjunct instructor basis, more widespread integration is clearly an issue for the academic side of the university. A clear desire to have the sustainability director work with both parts of the university is reflected by the six universities that have the sustainability coordinator report directly to the president, or jointly to the provost and physical plant, or to the VP for administration (to whom physical plant directors commonly report).

While at many universities the physical plant/facilities directors are among the leading proponents for sustainability, at some campuses these individuals may still be skeptical of sustainability efforts. In the face of such reluctance or outright resistance, locating the coordinator within the facilities department, where his or her suggestions and ideas can be constantly vetoed, may seem counterproductive. However, sustainability coordinators may be better positioned to overcome or minimize resistance by being a member of the facilities department and thus having the opportunity to demonstrate their abilities, trustworthiness, and value to the unit on a regular basis, than by being an outsider who is viewed as not understanding the challenges and constraints that facilities staff must operate under. The question of the optimal location of the sustainability coordinator position is deserving of future attention as the position continues to become more widespread.

Career Issues

Many sustainability coordinators have a background in environmental studies or in a closely related area, such as biology or geography. Others have backgrounds in architecture, engineering, or law. Most of the job descriptions examined indicated that training in some environmental field was either required or desired, but public administration, planning, and architecture were also mentioned. Generally only a bachelor's degree is required. While over one-third of the job descriptions examined expressed a preference for a graduate degree, only four actually required one.

Reflecting the newness of the position, only one-third of the job descriptions (which in most cases were written in conjunction with hiring the initial coordinator) indicated that over five years of work experience was required, with only two of the job descriptions requiring prior experience as a sustainability coordinator. Kester (2005) observes that there is some tendency to hire recent college graduates to serve as sustainability coordinators. In part, this reflects the fact that many sustainability coordinator positions have been created as a result of student advocacy, and that, frequently, one of the primary champions is offered the new position. It also reflects the relatively low salaries associated with the position. A salary survey conducted by AASHE's predecessor organization indicated that in 2005 the average salary for coordinators with 0–5 years of experience was $34,000 for those without graduate degrees and $41,500 for those with graduate degrees. While salaries were higher for those with more experience, at the time only two of the individuals responding to the survey had salaries above $80,000 (EFS West 2005). It is highly unlikely that recent graduates will have developed all of the skills described earlier to the degree needed for success. As a new profession, there is no clearly defined career path leading to or from the sustainability coordinator position. With no clear subsequent job option, sustainability coordinators who leave the position frequently leave the organization as well, thus depriving their successors of their expertise and advice.

Challenges

Harvard Green Campus Initiative director Leith Sharp, one of the most well-known sustainability coordinators, identifies a number of challenges facing sustainability coordinators. She argues that in order to bring about significant change in campus processes (e.g., increase recycling, decrease energy use, increase procurement of green materials, and design and construct green buildings), sustainability practitioners must also, and, to a large degree, must first, change campus institutional drivers. Necessary steps include making known the hidden upstream and downstream environmental impacts of campus operations, developing the capabilities of a learning organization within the campus, aligning the sustainability mission across basic university functions (teaching, research, and operations), and becoming an advocate for regulations that facilitate sustainability (Sharp 2005).

To accomplish these changes, Sharp argues that sustainability coordinators should conceive of their role as an entrepreneurial one, emphasizing strategic facilitation. As such, sustainability coordinators, whose background generally involves some aspect of environmental studies, must become organizationally astute change agents. They must understand their organizations and stakeholders, be able to energize them to take action, help generate ideas and assist in their evolution, discover ways to find solutions to objections and thereby minimize resistance, identify and secure resources, effectively communicate and manage perceptions of risk, and clearly document requests, activities, and successes (Sharp 2005).

Drawing upon Piaget's notion that a critical achievement in child development is gaining a sense of "object permanence" (a child learns that objects still exist even though they are not seen at the moment), Sharp argues that

> the core cognitive challenge in relation to achieving campus sustainability is to sustain a sense of [environmental] *impact permanence* (knowledge of the hidden impacts) in the presence of an immediate context that makes these impacts disappear from view. Without having the mental representational abilities to sustain a sense of *impact permanence* associated with choices and behavior, the individual is triggered to behave as if there are no impacts. We, like the child, behave as if the impacts no longer exist, simply because they are not in our sight.
>
> (Sharp 2005 [emphasis in the original])

The organizational and change skills suggested by Sharp as needed, together with many of the other skills noted earlier, pose quite a challenge for the effectiveness of sustainability coordinators. By and large, sustainability coordinators are individuals drawn to the position by their passion for the environment; they generally have training in some area of environmental science or environmental studies. Attention to organizational analysis and change is practically nonexistent in such disciplines. Given the relatively low pay of the positions and the lack of understanding of the full nature of the position by those who serve on search committees, it would not be surprising if most new sustainability coordinators were ill prepared to be effective in their position. Despite this fact, those sustainability coordinators with whom the authors have interacted tend to exhibit a high degree of excitement and report significant accomplishments. It appears that their passion for the goal of advancing sustainability motivates them to substantial on-the-job learning and skill development and application.

The Future

Sustainability coordinators expect that interest in campus sustainability will continue to grow, and that this interest will fuel continued growth in the number of sustainability coordinator positions. Despite the apparent significance of sustainability coordinators to campus sustainability efforts, the growth of this structural innovation has attracted practically no research attention. A recent article, reporting on the use of a Delphi technique to identify priorities for research in the area of higher education for sustainability, identified "institutional culture and organizational/governance structures" as being tied for the third most important areas of research (Wright 2007). Surprisingly, however, no mention was made about any aspect of the sustainability coordinator position within this area of research.

Walton and Galea (2005) suggest that there is much that academia can learn and apply from business when it comes to improving sustainability. However, it appears that academia is far ahead in the area of using sustainability coordinators to create and manage sustainability programs and initiatives. Can this innovation be diffused to the governmental sector, and even to business?

Sustainability Coordinators in the Public Sector

With over 38,000 units of local government in the United States (county and municipal governments) and another 13,500 school districts and hundreds of state agencies (Bureau of the Census 2002), all expending hundreds of billions of dollars annually, the role of sustainability coordinators is almost perfectly situated for government. Furthermore, the nature of government as a societal leader in intergenerational issues and as a steward of public goods places government entities at the forefront for considering such initiatives. However, the number of governmental entities currently employing sustainability coordinators is still quite low. In the absence of a systematic census of this position for government employment, it is impossible to say exactly how many exist in state and local government, but it would seem safe to say that anything covering more than 1 percent of public entities would be very optimistic.

In 2005 Portney reported that forty-two cities in the United States had initiated sustainability programs (Portney 2005). While none of Portney's work examined the position of sustainability coordinators in government, it is instructive to note that the author does not mention the position in any study of sustainability in government (Portney 2003, 2005). It is enlightening to use the Portney list of cities as a point of discussion concerning sustainability movements in American government. Six of the U.S.'s ten most populous cities have sustainability programs (New York, Los Angeles, Chicago, Phoenix, San Diego, and San Jose). Houston, Philadelphia, San Antonio, and Dallas lack such programs. Thus, roughly 60 percent of major cities (the numbers are maintained if you move out to the twenty largest cities) have sustainability programs. However, since only forty-two programs were identified, it is clear that as city size decreases, the likelihood of sustainability programs and sustainability coordinators drops quickly.

Sustainability has been sold at the local and state government level as a means of economic development and costs savings. Several studies have shown how local governments can better situate themselves in terms of future growth and economic vitality based on sustainability programs (Bartle and Leuenberger 2006; Leuenberger 2006; Stone and Dzuray 2005; Willis 2006). These authors have sought to make the case that sustainability has tangible long-term growth consequences for communities and that long-term viability and economic advantage will be gained through such activities.

At this point we will examine the sustainability programs of a few cities that currently employ sustainability coordinators and some of the job descriptions and structural forms associated with such jobs. It is worth noting that several cities in England, Canada (Vancouver, in particular), and Australia have very robust sustainability programs but this paper limits its examination to U.S. models.

One of the most successful and long-lived sustainability programs and coordinator positions is in Santa Monica, California. The program grew out of a mayoral task force on environmental issues convened in 1991. The Sustainable City Program (SCP) was launched in 1994 after several years of planning and preparation. The program has goals in four areas: resource conservation, transportation, pollution prevention and public health, and community and economic development. Eighteen different targets and indicators are attached to these areas and they are monitored and reported upon on a biannual basis (available at http://www.smepd.org). As an example, Objective One is, "Strive for the sustainable use of nonrenewable and limited resources such as fossil fuels, metals, minerals, and water," and the first policy goal under this objective is, "Maximize the energy efficiency of all new and existing buildings in the city." This goal has been partially achieved by Santa Monica as the city now reports that it purchases 100 percent of its energy needs from renewable sources and that all facilities have been retrofitted to improve energy efficiency and reduce costs.

This entire program comes under the purview of Dean Kubani, the sustainable city program coordinator. His job description is essentially to implement the SCP program in Santa Monica. Kubani holds degrees in geology and environmental geology and has studied architecture. Kubani describes his job as a consultant to other city divisions and departments. It is his job, along with his staff of seventeen, to assist other departments in their efforts to implement the SCP.

Andrew Watterson was hired by the City of Cleveland in 2004 as its first "sustainability programs manager" (he was previously employed in real estate development with a focus on historic restoration and green building). His job title is structured much like the one in Santa Monica in the sense that Mr. Watterson is to advise the city on projects relating to energy, buildings, fleet use, and purchase. The goals of the sustainable programs in Cleveland are (1) to save the city money and reduce its ecological footprint, (2) to use sustainable principles as a tool for economic development, and (3) to introduce sustainability principles to city employees through education. The program was initiated by a group of local environmental and sustainability organizations, and many of these individuals now oversee the program as members of the steering committee. The steering committee has ten members drawn from across the political spectrum of Cleveland.

An editorial comment from *The Steamboat Pilot* (2007) of Steamboat Springs, Colorado, may well depict the current status of sustainability coordinators. The city council of Steamboat adopted a Sustainability Management Plan in August 2006. In early 2007 the Green Team, a volunteer group that had spearheaded the plan, proposed hiring a sustainability manager for the city. It was estimated that the position would cost the city between $80,000 and 100,000 a year. The editorial appearing in the local paper supported the town council position that hiring such a person was not necessary at this time and should be reconsidered in the 2008 budget. The rationale read as follows:

> Don't misunderstand—we support the city's Green Team and efforts at sustainability. But sustainability is a not a government program and it's not a city department. It's simply a philosophical approach to governance.

This sentiment would seem to raise key questions about the future of sustainability and sustainability coordinators across government. First, will sustainability be seen as a philosophical approach worth taking, given the norms and values of government and government employees? Second, if the philosophical hurdle is passed, will the citizenry decide employees with specific skill sets are needed in order to accomplish this philosophical mandate or will this be achieved in some other fashion?

The examples cited here and numerous others point to an age-old question when it comes to organizing a new initiative. Should the program or activity be created from a top-down approach or from something more grassroots? The examples here show both can work when it comes to sustainability activities and the eventual hiring of a coordinator. Indeed Portney (2005) reached the same conclusion when he examined sustainable cities initiatives in his work. Given the number of state and local governmental units, it would seem we will see a dramatic increase in the number of sustainability programs and employees in government in the coming years, but just how those positions will be justified, compensated, and ultimately empowered is still to be determined.

Sustainability Coordinators in the Private Sector

An increasing number of corporations are adopting sustainability commitments as a piece of their mission, as one of their goals, or as a strategic initiative pertaining to a component of the firm. One indicator of this trend is the increasing number of firms that produce some sort of corporate citizenship report. Many of these companies file their reports on their websites to allow access of stakeholders to this information and also file the reports with public repositories of sustainability reports For example, Sustainability Reports has developed a portal for public sharing of corporate sustainability reports globally (http://www.sustainability-reports.com/), which archives and allows public access to these reports by any stakeholder. Another indictor is the number of organizations that are listing sustainability as one of the areas in which they are a consultant for corporations, such as the Institute for Sustainable Futures in Australia (ISF 2007).

At least three dimensions appear to influence the structures behind what is done, why it is done, and how it is done in corporate sustainability initiatives. These dimensions relate to the founding of the company, the amount of initiative integration throughout the firm, and the extent to which the company carries its sustainability initiatives beyond the boundaries of the firm. The first dimension is whether sustainability in some form was involved in the founding of the company. In some cases, such as Ben and Jerry's or Patagonia, sustainability concepts are, in whole or part, the motivation for founding the company. This motivation may not always mean there are immediate sustainability successes or innovations, but if the founders' values are influenced by sustainability, the core values of the firm are typically shaped by sustainability and eventually those values affect decisions of the corporation. In cases where sustainability becomes an important value after the firm has been established, the commitment to sustainability seems to become public most often when there is a leadership change to an individual who holds sustainability as a personal value. Two cases that illustrate this point are Ford Corporation under the leadership of Bill Ford and WalMart under the leadership of Lee Scott.

A corollary of the founding influence is whether the corporation's ownership is public or private. To a large extent, a private corporation has to answer only to its owners, usually in a relatively restrained and private arena, while a public corporation faces scrutiny in a much more visible arena and from a larger collection of stakeholders: its stockholders, the market, financial analysts, and often a large set of other potentially more vocal stakeholders who can focus and energize public opinion.

The second dimension is whether sustainability is an integrated focus permeating the firm or whether it is a compartmental focus of one segment of the firm. Patagonia illustrates both this dimension and the previous one. The very mission and raison d'être of the firm is "Build the best product, do no unnecessary harm, use business to inspire and implement solutions to the environmental crisis" (Patagonia 2007). Yvon Chouinard, owner of Patagonia, states that as a private firm Patagonia has the luxury of basing every business decision upon its mission with no pressure to meet shareholder's earning expectations (Little 2004). Thus, their commitment to sustainability permeates everything they do. More commonly, it seems that corporations initially focus sustainability on one aspect of their firm such as product development (Toyota and hybrid cars) or supply chain issues (McDonald's and beef purchases) or pollution (KLM and noise pollution/carbon dioxide generation). Some firms then evolve to a broader approach to sustainability (for example, McDonald's and Toyota), but others remain more focused on a key area (see, for example, KLM 2007). Remaining focused often happens when there is a sustainability issue within their business which is particularly visible and salient to stakeholders. Therefore, sustainability can be focused purely on a limited domain of issues, such as direct environmental issues like sustainable supplies of raw materials or products, or it can be applied as a general focus of the firm in the sense of looking for multiple and long-term ways to increase the firm's alignment with the needs of a sustainable world. One corollary to this dimension that may be essential to the success of sustainability emphases is the centrality and power of those accountable for sustainability efforts. While employee excitement can be a very heady source of energy and power, it can also be quite fickle if employees become involved in other areas or if key people leave the firm. Thus, dispersing sustainability efforts can be both good (involving most or all of the employees in sustainability awareness) and bad (no one has the power or network centrality to make significant changes occur without stepping out to become a change agent).

The third dimension is whether companies engage in activism beyond their stance on sustainability affecting their internal corporate activities. Many companies focus their sustainability activities on policies and practices to improve the sustainability of their own products and processes; other companies use their products and marketing as a means of environmental education and activism. Ben and Jerry's ice cream carries messages on the cartons about various social issues including environmental causes. British Petroleum (BP) ads encourage responsible use of fuels. Other organizations, even very small ones, engage their employees in volunteer activities that may include environmental cleanup actions (e.g., picking up trash along highways).

On the basis of these three dimensions, there are many examples of corporations engaging in sustainability activities using a wide range of structures. These structures may be voluntary or mandatory; they range from ad hoc arrangements to ones incorporated into the formal structure of the firm. The sustainability activities may be widely inclusive of all employees, the domain of sustainability specialists, or integrated with another aspect of the firm such as quality, strategy, or environmental compliance. Ben and Jerry's uses Green Teams that are voluntary groups of employees with an interest in environmental issues. Green Teams brainstorm ideas and projects that can move the facility or the overall firm toward sustainability (Ben and Jerry's 2006). These teams are not the sole source of environmental improvements within Ben and Jerry's, but they do encourage creativity and involve a broader set of employees in detecting areas for action or improvement.

Baxter Corporation used a Sustainability Steering Committee of executives to lead the incorporation of sustainability initiatives into its work activities and mission (Baxter Corp. 2007). Baxter's focus thus far has ranged from improving the efficiency of its use of natural resources, such as water, to reducing the environmental impacts of its medical products, and to making its headquarters facilities carbon neutral. While most of Baxter's focus is internal, on products, processes, and workplace issues, it has taken an advocacy stance on health-care-system change that it sees as an underpinning of overall social sustainability.

Toyota North American has an Environmental Working Group (EWG), which serves as the focus for its environmental activities at each of Toyota's facilities (Toyota 2007). This group reports to the Toyota North American Secretariat, which then reports to the North America Environmental Committee, which reports in turn to the North American Presidents Group. The EWG develops the five-year plan for environmental actions for Toyota and evaluates progress toward environmental goals across time. Unlike Ben and Jerry's volunteers, Toyota's example includes appointed people and, like Baxter, tends to include mostly higher-level executives.

BMW involved its stakeholders in its sustainability efforts by surveying a sample of them to understand how BMW was perceived and what areas of sustainability were seen as most important to be addressed (BMW 2007). BMW also examined its competitors to see what areas of focus were competitively key in its competitor's sustainability efforts, in essence identifying how sustainability was altering the forces of rivalry in the high-end auto market globally. Therefore, BMW has wedded sustainability to corporate reputation and attempts to instill sustainability in all management decisions. Thus, it charges all of its employees with moving BMW toward more sustainable practice in all the venues in which they operate. BMW encourages cross-functional discussion to seek the most sustainable practices possible in each circumstance.

Office Depot has adopted the slogan of "Buying Green, Being Green, Selling Green" as its mantra in efforts to be more sustainable. Its logic is to develop a strong environmental policy, define responsibility, and then implement an environmental management system (Office Depot 2007). Office Depot has also hired Yalmaz Siddiqui as its environmental strategy advisor. Among the companies examined, this case is the clearest one of someone being hired with specific sustainability interests and education to guide the corporate activities and commitments. Mr. Siddiqui has both an educational background and consulting experience in sustainability and environmental issues. Thus, Office Depot has moved to bring in an external expert instead of using already established corporate executives, training people in situ, or relying upon the personal environmental motivations of its employees. In addition to improving its own sustainable practices, Office Depot has involved stakeholders in its sustainability efforts in a manner similar to BMW's efforts. It has begun awarding a "Green Customer Award" to its five major customers who have made significant investments in greener products from Office Depot. In addition, Office Depot continues to help its customers improve their sustainability by providing green choices and educating them about green choices.

Conclusion

This chapter provides an introduction to a new organizational position, the sustainability coordinator, and suggests that the sustainability coordinator represents a valuable structural innovation for managing sustainability. This role is becoming increasingly common on college and university campuses. It is also beginning to appear in municipal governments. As of yet, there is little evidence that this innovation has diffused to the corporate sector. This lack of rapid diffusion is perhaps not surprising, for several reasons.

First, the nature of the output of universities, governments, and businesses is significantly different. The environmental impacts of the products, production processes, packaging, and, to some degree, the distribution methods of business have drawn substantial attention for nearly four decades. Given these visible impacts, the impacts of organizations support activities—for example, water, materials, and energy use—have drawn less attention. As such, companies that are still struggling to meet public demands for greening their products and reducing pollution from their factories have less incentive to attend to the sustainability of their office and support activities. Universities and governments, however, produce relatively little in the way of products. They are more like service industries than manufacturing industries. Thus, questions regarding their environmental impacts quickly lead to a focus on the internal support activities that generate the services they produce. This perspective suggests that we might expect to see the emergence of sustainability coordinators in the corporate sector first in service industries.

A second reason is the difference in both awareness of and concern for sustainability by core stakeholders. All three sectors, universities, organizations and government, have employees, and in all three sectors there is anecdotal evidence that employee concern has driven the establishment of sustainability initiatives, particularly the creation of recycling programs. But the other core stakeholders of the three sectors vary significantly. The stockholders and customers of business and the citizen constituents of governments have far less day-to-day intensive interaction with, and thus awareness of, the activities of these organizations than do students in universities. In addition, the degree of communication and interaction between students and one group of employees—faculty—facilitates the mutual transmission and reinforcement of concern for sustainability. Thus, students and faculty are more likely to be in a position to influence their universities to make a sustainability commitment than are the stakeholders of businesses and governments. However, the situation in businesses and governments may be changing. If public concern for issues of environmental sustainability increases significantly, as attention to global warming suggests it may, then we may expect to see increased interest in and attention to organizations' efforts at making internal operations more consistent with the need for sustainability. Such attention may be considerably enhanced if organizational transparency increases, thereby allowing stakeholders to observe more easily the practices an organization is engaging in.

Such a stakeholder-concern-driven analysis might suggest that sustainability coordinators will emerge more rapidly in government than in business if citizens are more similar to students than to customers in feeling able to express indignation that an organization is not attempting to improve its sustainability.

Given the predominantly economic nature of the stockholder-corporation relationship, we would not expect stockholders to exert much pressure on corporate management for internal sustainability, unless such action becomes widely seen as a means for either reducing costs or increasing profits. This change of perception may be starting to occur, particularly on the cost reduction side, given the rise in energy prices. Attention to energy efficiency for economic reasons may lead to a greater awareness of opportunities to save money while reducing environmental impacts in a variety of domains beyond the energy one. Increasing awareness of such possibilities may offer opportunities for environmentally concerned employees to make suggestions for other win-win prosustainability practices and increase companies' willingness to consider allocating specific responsibilities for identifying cost-saving and sustainability-enhancing practices. To the degree that employees have become aware of sustainability efforts as students, such an evolution may be hastened. Thus, we might expect to see the emergence of sustainability coordinators in business sooner among those companies with a relatively young and well-educated work force, such as entrepreneurial start-ups and technology sector companies.

Whichever of these patterns might drive the diffusion of this structural innovation, we believe that the sustainability coordinator position will continue to grow rapidly in the university sector and at an increasing rate in the governmental and corporate sectors. As it does, we believe that the chances of successfully moving toward a more sustainable future will be significantly enhanced.

Acknowledgments

The authors wish to thank MBA students Tyler Sayers and Pamela Rands for research and editorial assistance.

References

AASHE: Association for the Advancement of Sustainability in Higher Education. 2007a. Association for the Advancement of Sustainability in Higher Education. http://www.aashe.org (accessed March 1, 2008).

AASHE: Association for the Advancement of Sustainability in Higher Education. 2007b. Campus sustainability officers. http://www.aashe.org/resources/sust_professionals.php (accessed March 1, 2008).

AASHE: Association for the Advancement of Sustainability in Higher Education. 2007c. Directory of campus sustainability/environmental officers. http://www.aashe.org/resources/sodirectory.php (accessed March 1, 2008).

Akel, M. 2006. A greener attitude. *University Business.* http://www2.universitybusiness.com/viewarticle.aspx?articleid=549 (accessed March 1, 2008).

Bartle, J., and Leuenberger, D. 2006. The idea of sustainable development in public administration. *Public Works Management & Policy* 10 (3): 191–194.

Bartlett, P. F. 2004. No longer waiting for someone else to do it: A tale of reluctant leadership. In *Sustainability on campus: Stories and strategies for change,* ed. P. F. Bartlett and G. W. Chase, pp. 67–87). Cambridge, MA: MIT Press.

Baxter Corporation. 2007. Baxter Corporation 2006 Sustainability Report. sustainability.baxter.com (accessed March 1, 2008).

Ben and Jerry's. 2006. Greening our workforce segment of social responsibility report for 2006. http://www.benjerry.com/our_company/about_us/social_mission/social_audits/2005_sear/sear05_6.6.cfm (accessed March 1, 2008).

Benderoff, E. 2007, August 23. Class of 2011 encouraged to adopt greener lifestyle. *Chicago Tribune.* www.chicagotribune.com/news/nationworld/chithu_ greenschoolaug23,1,921406 (accessed February 15, 2008).

BMW. 2007. Sustainable Value Report 2007/2008 of Bayerisch Motoren Werke. http://www.bmwgroup.com/responsibility/ (accessed March 1, 2008).

BSU: Ball State University. 2007. Greening of the campus VII. http://www.bsu. edu/greening (accessed March 1, 2008).

Bureau of the Census. 2002. *2002 census of governments.* Washington, D.C.: U.S. Government Printing Office.

Campus Consortium for Environmental Excellence/Environmental Association for Universities and Colleges 2006. A practical guide to hiring a sustainability professional for universities and colleges. http://www.c2e2.org/sustainability_guide. pdf (accessed February 15, 2008).

Clarke, A. 2004. *Campus environmental management systems: Dalhousie University as a case study.* Paper presented at the annual meeting of the Academy of Management.

Creighton, S. H. 1998. *Greening the ivory tower.* Cambridge, MA: MIT Press.

Dautremont-Smith, J. 2007. Telephone interview with Gordon Rands, 9/12/07.

Deutsch, C. 2007, July 13. College leaders press for carbon neutrality. *New York Times.* http://www.nytimes.com/2007/06/13/education/13green.html?_r=1& oref=slogin. (accessed February 15, 2008).

Eastman, J. 2007, August 23. Green goes to school. *Los Angeles Times.* http://www. latimes.com/features/printedition/home/la-hm-green23aug23,0,5494975.story. (accessed August 25, 2007).

EFS West. 2005. EFS-West salary survey of sustainability and environmental coordinators. http://www.aashe.org/resources/pdf/EFSSalarySurvey2005.pdf (accessed March 1, 2008).

Grist: Environmental News and Commentary. 2007. 15 green colleges and universities. http://www.grist.org/news/maindish/2007/08/10/colleges/index.html (accessed March 1, 2008).

Hails, C., Loh, J., and Goldfinger, S. 2006. The Living Planet Report, 2006. http:// assets.panda.org/downloads/living_planet_report.pdf (accessed January 15, 2007).

Hattam, J. 2007. Go big green. *Sierra,* November/December. http://www.sierraclub. org/sierra/200711/coolschools (accessed March 1, 2008).

Heinz Family Foundation, The. 1995. Blueprint for a green campus: The Campus Earth Summit Initiatives for higher education. http://www.heinzfamily.org/pdfs/ Blueprint-For-Green-Campus.pdf (accessed February 15, 2008).

ISF: Institute for Sustainable Futures. 2007. What we do. http://www.isf.uts.edu. au/whatwedo/proj_ICS.html (accessed March 1, 2008).

Kester, C. 2005. Recent graduates as sustainability coordinators: Challenges and opportunities. In *Conference proceedings: Greening the campus VI: Extending connections,* ed. R. Koester. CD. Muncie, IN: Ball State University.

KLM. 2007. 2006/2007 A year in perspective. Online sustainability report. http:// publieksverslag.klm.com/en/index.html (accessed March 1, 2008).

Leuenberger, D. 2006. Sustainable development in public administration: A match with practice? *Public Works Management & Policy* 10 (3): 185–201.

Little, A. G. 2004. Don't get mad, get Yvon. http://www.grist.org/news/maindish/2004/10/22/little-chouinard/index.html (accessed March 1, 2008).

Meadows, D. H., Meadows, D. L., and Randers, J. 1992. *Beyond the limits: Global collapse or a sustainable future*. London: Earthscan Publications.

Miller, R. A., Umashankar, R., and Mella, G. 2005. Greening UConn: Implementing the university's sustainability vision. In *Conference proceedings: Greening the campus VI: Extending connections*, ed. R. Koester. CD. Muncie, IN: Ball State University.

National Wildlife Federation. 2007. Campus ecology. http://www.nwf.org/campusecology (accessed March 1, 2008).

Nicolaides, A. 2006. The implementation of environmental management towards sustainable universities and education for sustainable development as an ethical imperative. *International Journal for Sustainability in Higher Education* 7 (4): 414–424.

Office Depot. 2007. Office Depot n Citizenship Report 2007. www.officedepot.com/corporatecitizenship (accessed March 1, 2008).

Orr, D. W. 1990. What is education for? *Environmental Professional* 12 (3): 351–355.

Orr, D. W. 1992. *Ecological literacy: Education and the transition to a postmodern world*. Albany, NY: SUNY Press.

Orr, D. W. 1994. *Earth in mind: On education, environment, and the human prospect*. Washington, DC: Island Press.

Patagonia. 2007. Our Reason for being segment. http://www.patagonia.com (accessed March 1, 2008).

Portney, K. 2003. *Taking sustainable cities seriously: Economic development, quality of life, and the environment in American cities*. Cambridge, MA: MIT Press.

Portney, K. 2005. Civic engagement and sustainable cities in the United States. *Public Administration Review* 65 (5): 579–591.

Price, T. J. 2005. Preaching what we practice: Experiences from implementing ISO 14001 at the University of Glamorgan. *International Journal of Sustainability in Higher Education* 6 (2): 161–178.

Rands, G., Ribbens, B., and Bingham, J. 2006. *Campus sustainability officers: The emergence and maintenance of a new career*. Paper presented at the Greening of Industry Network conference.

Second Nature. 2007. Education for sustainability. http://www.secondnature.org/efs/efs_organizations.htm (accessed March 1, 2008).

Sharp, L. 2005. Campus sustainability practitioners: Challenges for a new profession. In *Conference proceedings: Greening the campus VI: Extending connections*, ed. R. Koester. CD. Muncie, IN: Ball State University.

Shriberg, M. 2002. Institutional assessment tools for sustainability in higher education: Strengths, weaknesses, and implications for practice and theory. *Higher Education Policy* 15: 153–167.

Shrivastava, P. 1995. The role of corporations in achieving ecological sustainability. *Academy of Management Review* 20 (4): 936–960.

Skov, J. 2004. *The challenges of campus greening: An external consultant's perspective.* Paper presented at the annual meeting of the Academy of Management.

Starik, M., and Carroll, A. B. 1992. Strategic environmental management: Business as if the earth really mattered. In *Contemporary issues in the business environment,* ed. D. Ludwig and K. Paul, pp. 143–169). Lewiston, PA: Edwin Mellen Press.

Starik, M., and Rands, G. P. 1995. Weaving an integrated web: Multilevel and multisystem perspectives of ecologically sustainable organizations. *Academy of Management Review* 20: 908–935.

Starik, M., Schaeffer, T. N., Berman, P., and Hazelwood, A. 2002. Initial environmental project characterizations of four U.S. universities. *International Journal of Sustainability in Higher Education* 3 (4): 335–345.

Steamboat Pilot & Today, The. 2007, May 20. Our View: Sustainability doesn't need a coordinator, http://www.steamboatpilot.com/news/2007/may/20/our_view_sustainability_doesnt_need_coordinator/ (accessed September 15, 2007).

Steptoe, S. 2007, August 10. Getting schools to think and act green. *Time.* http://www.time.com/time/specials/2007/article/0,28804,1651473_1651472_1652067,00.html (accessed August 15, 2007).

Stone, S., and Dzuray, E. 2005. Achieving environmental sustainability in government operations. *Public Manager* 34 (4): 22–26.

Sustainable Endowments Institute 2007. College sustainability report card, 2008. http://www.endowmentinstitute.org/sustainability/CollegeSustainabilityReportCard2008.pdf (accessed February 15, 2008).

Toyota 2007. Toyota North America Environmental Report 2006. http://a230.g.akamai.net/7/230/2320/v001/toyota.download.akamai.com/2320/toyota/media/about/2006envrep.pdf (accessed March 1, 2008).

Uhl, C. 2004. Process and practice: Creating the sustainable university. In *Sustainability on campus: Stories and strategies for change,* ed. P. F. Bartlett and G. W. Chase, pp. 29–47. Cambridge, MA: MIT Press.

ULSF: University Leaders for a Sustainable Future. 2007. Talloires declaration. http://www.ulsf.org/talloires_declaration.html (accessed March 1, 2008).

Walton, S. V., and Galea, C. E. 2005. Some considerations for applying business sustainability practices to campus environmental challenges. *International Journal of Sustainability in Higher Education* 6 (2): 147–160.

Willis, M. 2006. Sustainability: The issue of our age, and a concern for local government. *Public Management* 88 (7): 8–12.

Wright, T. S. A. 2007. Developing research priorities with a coalition of higher education for sustainability experts. *International Journal of Sustainability in Higher Education* 8 (1): 34–43.

Notes on Contributors

Douglas E. Allen received his PhD in marketing from the Pennsylvania State University. He is currently an associate professor of management at Bucknell University, Lewisburg, PA. Doug's teaching and research interests are in the area of consumer culture theory, focusing on the sociohistorically embedded nature of consumer practices. In addition to publishing articles on consumer culture in a wide array of journals, he has developed courses that explore trends in sustainable consumption. He is also a member of the Institute for Leadership in Technology and Management that focuses extensively on themes related to sustainability.

Geoff Archer is on the faculty of the College of Business at Oregon State University. He is also a PhD candidate in entrepreneurship and ethics at the Darden Graduate School of Business at the University of Virginia. The working title of his dissertation is "'Base of the Pyramid' Schemes: Who Cares About the Natural Environment in the Microfinance Process?" This multimethod research hopes to explain the existence of and the drivers behind a handful of microfinanced initiatives in which entrepreneurs have deployed "cleantech" solutions (i.e., solar panels in a remote Argentinean village). Receiving his MBA from Cornell University, Geoff was a member of the first class of Park Leadership Fellows, focusing his studies on entrepreneurship and international business. Geoff holds a bachelor's degree in public policy studies and a master of environmental management from Duke University. His concentration at Duke was resource economics and policy, which culminated in a thesis, "Corporate Communication of Environmental Performance," paying particular attention to start-ups.

David R. Connelly earned his PhD in 2005 from the Utah Valley University. David currently teaches public administration courses as an associate professor in the Political Science Department at Western Illinois University. His research interests include leadership in collaborative environments, information technology issues in the public sector, and public sector management reform.

Frank G. A. de Bakker is an assistant professor of strategic management at the Faculty of Social Sciences in the VU University Amsterdam, the Netherlands. Corporate social responsibility, social movement organizations, activism, institutional change, and stakeholder management are important elements of his research, which has been published in journals such as the *Academy of Management Review, Business & Society, European Management Journal,* and *Business Strategy and the Environment.*

Frank Figge holds a PhD from the University of Basel (Switzerland). Prior to joining academia Frank worked for asset managers in the field of socially responsible investment. Frank currently holds the chair of management and sustainability at Queen's University Management School (UK). Frank serves on several advisory boards. His research focuses on value-based sustainability management, the economics and management of diversity, strategic sustainability management, and stakeholder management.

Tobias Hahn holds a masters degree in environmental science and a PhD in economics and social science on stakeholder behavior and management. He is currently an associate professor for corporate sustainability, corporate social responsibility, and environmental management at Euromed Marseille Business School in France. Prior to this he worked as a senior researcher at the Berlin-based Institute for Futures Studies and Technology Assessment (Germany). His major research interests are in the field of management and measurement of corporate sustainability as well as in analyzing stakeholder behavior. Tobias serves as a board member of academic journals. In recent years his research and publications have focused mainly on the development and application of value-based approaches to corporate sustainability.

Tammy Bunn Hiller holds a PhD in organizational behavior from the University of North Carolina at Chapel Hill. She is an associate professor of management at Bucknell University, Lewisburg, PA. She teaches about organizing for justice and social change, both in a classroom setting and through leading intensive service-learning alternative break trips to Nicaragua and the U.S. Gulf Coast. She also writes about service-learning and experiential-learning pedagogy related to teaching about sustainability.

Nicole A. M. Horstman (MSc) graduated from the VU University Amsterdam, and the University of Twente, the Netherlands.

Andrea Larson is Associate Professor at the Darden Graduate School of Business Administration where she teaches electives on entrepreneurship, innovation, and sustainability. She holds a joint PhD from Harvard University's Business School and Harvard Graduate School of Arts and Sciences. Her publications have appeared in journals including *Administrative Science Quarterly, The Journal of Business Venturing, Entrepreneurship Theory and Practice,* and the international operations journal *Interfaces.* Cases and other curriculum materials she has developed on sustainability and entrepreneurial innovation are available through Darden Publishing.

Mats A. Lundqvist is director and cofounder of Chalmers School of Entrepreneurship. The school operates with the dual objectives of developing entrepreneurs and new ventures by arranging partnerships with its students and inventors from Chalmers and the Gothenburg region, thereby constituting a new mechanism for increasing the commercialization of new technology. As a direct result, more than thirty high-tech start-ups have been initiated since 1997. As associate professor in innovation management, Mats carries out research, teaching, and collaborative development work in the areas of entrepreneurship and intellectual property management at the Center for Intellectual Property Studies. Mats is engaged in the board of several start-ups and in the steering of national-, regional-, and university-level programs concerned with commercialization of research.

Enno Masurel is a full professor in sustainable entrepreneurship at the VU Centre for Enterpreneurship of the VU University Amsterdam, the Netherlands. He studied business administration at the VU University Amsterdam from 1979 to1986. After obtaining his master's degree, he joined the Economic and Social Institute of VU University Amsterdam as a researcher. He obtained his PhD in 1993 with a thesis entitled "Small Business Collaboration in the Dutch Retail Sector" and then became head of the Business Administration Department of the Economic and Social Institute. In March 2004 he was appointed professor in sustainable entrepreneurship. His main research focus is entrepreneurship and small and medium-sized enterprises (SMEs), with special reference to innovation. Enno is a member of the Editorial Review Board of the *Journal of Small Business Management.* He has attended many international seminars and published in a number of international journals. In a recent analysis by *Technovation* concerning publications in the field of entrepreneurship, he was ranked in the Dutch top three articles and in the world top eight.

Karen Williams Middleton is a doctoral candidate at the Division of Management of Organizational Renewal and Entrepreneurship (MORE) at Chalmers University of Technology, in Gothenburg, Sweden. Karen holds a bachelor's degree in civil engineering from Tufts University and an MBA (with honors) from Babson College. Since 2004, Karen has been working with the Chalmers School of Entrepreneurship (and its sister program GIBBS, 2005) in areas including leadership and entrepreneurial development. She has also been engaged in entrepreneurial development projects, from the regional to the EU level, including a four-year-long project focused on entrepreneurial attraction and competence development at the regional level, and a one-and-a-half-year-long EU project focused on university research commercialization and mobility. Karen is currently engaged in project proposals regarding interregional innovation development, regional entrepreneurial culture development, and societal entrepreneurship from a Swedish perspective.

Laurie P. Milton earned her PhD from the University of Texas at Austin, her MSc from the University of Alberta, and her MBA from the University of Calgary. Laurie is an associate professor at the Haskayne School of Business, University of Calgary, and a Research Professor at the Richard Ivey School of Business, University of Western Ontario. Her research focuses on cooperation, collaboration, and identity (especially in interdependent contexts that involve knowledge sharing and development). Prior to completing her doctorate, Laurie held a series of management positions in the Strategic Planning and Research Secretariat of the Alberta Department of Housing and in the Legislative Assembly of Alberta. She has a strong managerial background in public policy research, program design, and evaluation. Laurie regularly presents her research at management and engineering conferences and publishes in leading academic journals.

Gordon Rands received his PhD in 1994 from the University of Minnesota. Gordon teaches courses on business ethics, business and the natural environment, organizational behavior, and organizational change in the Management Department at Western Illinois University, Macomb, IL, where he is an associate professor. His research focuses on various topics relating to organizations and the natural environment, including the topics of ecologically sustainable organizations, societies, and governments. He is a cofounder and past officer of the Organizations and Natural Environment (ONE) division of the Academy of Management and an active member of the Greening of Industry Network. He also serves on the sustainability committee and other environmental initiatives of his university.

Barbara A. Ribbens received her PhD in 1994 from the University of Connecticut. Barbara is an associate professor in the Management Department at Western Illinois University, where she currently teaches strategic management, career management, and decision-making courses. Her research interests include careers and decision making as they affect individuals and organizations. She has worked on a team that developed a framework for sustainable societies, sustainability issues in the Marshall Islands, and moral choices regarding Tuvalu. She is currently working on two projects: careers of sustainability coordinators and environmental attitudes among Marshallese immigrants to the United States.

Paul Shrivastava received his PhD from the University of Pittsburgh and currently holds the Howard I. Scott Chair in Management at Bucknell University, where he teaches courses in strategic management. He has published fifteen books and over one hundred articles. Paul has served on the boards of ten leading management journals and received a Fulbright Senior Scholar Award. He has consulted with leading corporations and founded the nonprofit Industrial Crisis Institute; *Organization and Environment*, a journal published by Sage Publications; and eSocrates, Inc., an online training software company. Before joining Bucknell, he was a tenured associate professor at New York University's Stern School of Business. Dr. Shrivastava serves on the board of trustees of DeSales University. He is coorganizer of the Steelman Triathlon and does a World Tango Music radio show on WVBU 90.5FM, Lewisburg, PA.

Joseph R. Sprangel Jr. has an MBA from Spring Arbor University and a BBA from Eastern Michigan University. He is an instructor of operations management for Ithaca College and is focused on sustainable development practices. He brings to the classroom twenty-eight years of industry experience, having been involved in the implementation of lean operations leading to waste elimination in the areas of administration, R&D, design, process development, production, and supply chain management. He is a DBA student at Lawrence Technological University. His research area of interest is development of a framework to move an organization from status quo to one of sustainability.

Jacqueline M. Stavros earned a doctorate in management at Case Western Reserve University. Her dissertation topic was "Capacity Building Using an Appreciative Approach: A Relational Process of Building Your Organization's Future." Her MBA is in international business from Michigan State University and her BA is from Wayne State University. Jackie is a professor for the College of Management, Lawrence Technological University, with twenty years of strategic planning experience. She teaches and integrates Appreciative Inquiry (AI) in her coursework: "Leading Organizational Change," "Strategic Management," "Organization Development," and

"Leadership." Jackie works with teams, divisions, and organizations to facilitate strategic change in a variety of industries. She has coauthored books, book chapters, and articles on AI and SOAR.

James A. F. Stoner is professor of management systems and chairholder of the James A. F. Stoner Chair in Global Sustainability at Fordham University's Graduate School of Business. Jim earned his BS in engineering science at Antioch College and his MS and PhD in industrial management at MIT. He has published articles in such journals and periodicals as *Academy of Management Review, Harvard Business Review, Journal of Experimental Social Psychology, Journal of Development Studies, Personnel Psychology*, etc., and has authored, coauthored, and coedited around twenty or so books and monographs including *Management* (first to sixth editions); *Fundamentals of Financial Managing* (first to second editions); and *Modern Financial Managing: Continuity and Change* (first to third editions). He has consulted with a wide variety of business organizations, received a number of teaching awards, and has taught in many executive development, MBA, and executive MBA programs, including the ones in Ethiopia, Iran, Ireland, Japan, and Siberia. He is a past chair of the Management Education and Development Division of the Academy of Management and past board member of the Organizational Behavior Teaching Society. The Stoner Chair was endowed in perpetuity in Jim's name at Fordham by a grant from Robert and Brent Martini, Jim's clients, very good friends, and in Brent's case, former star student.

Patricia P. van Hemert is currently working on her PhD on innovation, entrepreneurship, and economic growth. She has a background in European studies (University of Amsterdam). In 1998–1999 she spent a year in Ireland, where she studied history and politics at the University College, Dublin. At the moment, Patricia is working as a researcher at the VU Center for Entrepreneurship (VU University Amsterdam). For the centre, she participates in two European Commission Sixth Framework Programs—"Dynamic Regions in a Knowledge-Driven Global Economy: Lessons and Policy Implications for the EU (DYNREG)," and "Social Sciences and Humanities Futures (SSH Futures)." She has published in a number of international journals, such as, *International Journal of Foresight and Innovation Policy* and *Studies in Regional Science*.

Charles Wankel is associate professor of management at St. John's University, New York. He received his doctorate from New York University. Dr. Wankel has authored and edited many books on management education, the alleviation of poverty through business strategy, and sustainability. He is the leading founder and director of scholarly virtual communities for management professors, including the Organizations and Natural Environment Division of the Academy of Management's ONE-L, currently directing eight such communities with thousands of participants in more than seventy nations. He has taught in Lithuania at the Kaunas University of Technology (Fulbright Fellowship) and at the University of Vilnius (United Nations Development Program and Soros Foundation funding). Invited lectures include one to an Association of MBAs and EGADE sponsored on in 2008 in Mexico City. His corporate management development program clients include McDonald's Corporation's Hamburger University and IBM Learning Services and his pro bono consulting assignments include reengineering and total quality management programs for the Lithuanian National Postal Service.

Mark White is associate professor of commerce at the University of Virginia's McIntire School of Commerce, where he teaches corporate finance, international finance, and courses examining business' relationship with the natural environment. He holds advanced degrees in both finance and ecology. His current research focuses on the use of ecological capital, a topic melding his expertise in financial modeling with his interests in environmental conservation. Professor White is actively engaged in the identification and evaluation of innovative business strategies for achieving a sustainable society. He currently holds a visiting professorship in environmental economics at the Technische Universitaet, Dresden.

Jeffrey G. York is a research assistant at the Batten Institute at the University of Virginia's Darden School of Business and is a PhD candidate in entrepreneurship, business ethics, and strategy. He holds a bachelor's degree in journalism from the University of Georgia and an MBA from the University of Tennessee, where he focused on new venture analysis. His current work is focused on studying the nexus of environmental opportunities with entrepreneurial solutions. Prior to beginning his PhD, Jeff was a director of marketing operations at Capital One, where he led multiple strategic planning and new business development projects. Jeff's past work experience includes consulting for business incubators, working as a photojournalist, and leading the start-up of a white-water rafting company.

Index

Breinigsville, PA USA
22 February 2010
232903BV00001B/3/P